PRESENTING DIGITAL CASH

PRESENTING DIGITAL CASH

by Seth Godin

PUBLISHING

201 West 103rd Street
Indianapolis, Indiana 46290

Copyright © 1995 by Sams.net Publishing

FIRST EDITION

All rights reserved. No part of this book shall be reproduced, stored in a retrieval system, or transmitted by any means, electronic, mechanical, photocopying, recording, or otherwise, without written permission from the publisher. No patent liability is assumed with respect to the use of the information contained herein. Although every precaution has been taken in the preparation of this book, the publisher and author assume no responsibility for errors or omissions. Neither is any liability assumed for damages resulting from the use of the information contained herein. For information, address Sams.net Publishing, 201 W. 103rd St., Indianapolis, IN 46290.

International Standard Book Number: 1-57521-062-2

Library of Congress Catalog Card Number: 95-71695

98 97 96 95   4 3 2 1

Interpretation of the printing code: the rightmost double-digit number is the year of the book's printing; the rightmost single-digit, the number of the book's printing. For example, a printing code of 95-1 shows that the first printing of the book occurred in 1995.

Printed in the United States

Trademarks

All terms mentioned in the book that are known to be trademarks or service marks have been appropriately capitalized. Sams.net Publishing cannot attest to the accuracy of this information. Use of a term in this book should not be regarded as affecting the validity of any trademark or service mark.

Acknowledgments

Thanks to George Bond and Fran Hatton for the help and support that made this book a reality. George possesses a rare understanding of what the future holds.

Thanks also to John Mello, who worked hard to convert our first outlines and drafts into a workable manuscript, and to Jolanta Benal and Susan Meigs for their editorial input.

This book wouldn't exist without the insight and effort of Julie Maner. She designed the layout, trafficked the manuscript and dealt with a thousand details. Thank you Julie.

This book wouldn't exist without the insight and effort of Julie Maner. She designed the layout, trafficked the manuscript and dealt with a thousand details. Thank you Julie.

Thanks to Sidney Short and CJ Anastasio who put in countless hours on the technical layout and thanks to the rest of the Seth Godin Productions staff who contributed to this book: Amy Winger, Lisa DiMona, Anthony Schneider, Karen Watts, Robin Dellabough, and Malcolm Faulds.

This book is dedicated to Jay Conrad Levinson, guru, partner and friend.

About the Author

Seth Godin is the president of Seth Godin Productions, a rapidly growing creator and packager of information. He has written or edited more than 75 books including *The Information Please Business Almanac*, *Business Rules of Thumb*, *eMarketing*, *The Internet White Pages*, *Email Addresses of the Rich and Famous*, *Wisdom, Inc.*, *The Guerrilla Marketing Handbook*, and *Guerrilla Marketing for the Home-Based Business*. His work has been featured in *Fortune*, *Business Week*, *Rolling Stone*, *The Wall Street Journal*, *The New York Times*, and other publications.

Table of Contents

The Great Simoleon Caper, a story by Neal Stephenson 1

Shifting Paradigms 15

The Internet 21

One to One 35

Cryptology 45

Digital Cash 62

Mondex 84

DigiCash 99

CyberCash 114

First Virtual 125

Open Market 138

More Players 156

Global Implications 175

The Future 187

Getting Involved 197

Appendix 221
    Interview with Tim Jones of Mondex
    Interview with David Chaum of DigiCash
    Interview with Dan Schutzer of Citibank
    Interview with Don Gleason of Smart Card Enterprise
    Interview with Shikhar Ghosh of Open Market
    Company Contact Information

THE GREAT SI-MO-LE-ON CAPER

Neil Stephenson is the author of Snow Crash *and* The Diamond Age.

Hard to imagine a less attractive life-style for a young man just out of college than going back to Bismarck to live with his parents unless it's living with his brother in the suburbs of Chicago, which, naturally, is what I did. Mom at least bakes a mean cherry pie. Joe, on the other hand, got me into a permanent emotional headlock and found some way, every day, to give me psychic noogies. For example, there was the day he gave me the job of figuring out how many jelly beans it would take to fill up Soldier Field.

Let us stipulate that it's all my fault; Joe would want me to be clear on that point. Just as he was always good with people, I was always good with numbers. As Joe tells me at least once a week, I should have studied engineering. Drifted between majors instead, ended up with a major in math and a minor in art–just about the worst thing you can put on a job app.

Joe, on the other hand, went into the ad game. When the Internet and optical fiber and HDTV and digital cash all came together and turned into what we now call the Metaverse, most of the big ad agencies got hammered–because in the Metaverse, you can actually whip out a gun and blow the Energizer Bunny's head off, and a lot of people did. Joe borrowed 10,000 bucks from Mom and Dad and started this clever young ad agency. If you've spent any time crawling the Metaverse, you've seen his work—and it's seen you, and talked to you, and followed you around.

Mom and Dad stayed in their same little house in Bismarck, North Dakota. None of their neighbors guessed that if they cashed in their stock in Joe's agency, they'd be worth about $20 million. I nagged them to diversify their portfolio—you know, buy a bushel basket of Krugerrands and bury them in the backyard, or maybe put a few million into a mutual fund. But Mom and Dad felt this would be a no-confidence vote in Joe. "It'd be," Dad said, "like showing up for your kid's piano recital with a Walkman."

Joe comes home one January evening with a magnum of champagne. After giving me the obligatory hazing about whether I'm old enough to drink, he pours me a glass. He's already banished his two sons to the Home Theater. They have cranked up the set-top box they got for Christmas. Patch this baby into your HDTV, and you can cruise the Metaverse, wander the Web and choose from among several user-friendly operating systems, each one rife with automatic help systems, customer-service hot lines and intelligent agents. The theater's subwoofer caused our silverware to buzz around like sheet-metal hockey players, and amplified explosions knock swirling nebulas of tiny bubbles loose from the insides of our champagne glasses. Those low frequencies must penetrate the young brain somehow, coming in under kids' media-hip radar and injecting the edfotainucational muchomedia bitstream direct into their cerebral cortices.

"Hauled down a mother of an account today," Joe explains. "We hype cars. We hype computers. We hype athletic shoes. But as of three hours ago, we are hyping a currency."

"What?" says his wife Anne.

"Y'know, like dollars or yen. Except this is a new currency."

"From which country?" I ask. This is like offering lox to a dog: I've given Joe the chance to enlighten his feckless bro. He hammers back half a flute of Dom Perignon and shifts into full-on Pitch Mode.

"Forget about countries," he says. "We're talking Simoleons—the smart, hip new currency of the Metaverse."

"Is this like E-money?" Anne asks.

"We've been doing E-money for e-ons, ever since automated-teller machines." Joe says, with just the right edge of scorn. "Nowadays we can use it to go shopping in the Metaverse. But it's still in U.S. dollars. Smart people are looking for something better."

That was for me. I graduated college with a thousand bucks in savings.

With inflation at 10% and rising, that buys a lot fewer Leinenkugels than it did a year ago.

"The government's never going to get its act together on the budget," Joe says. "It can't. Inflation will just get worse. People will put their money elsewhere."

"Inflation would have to get pretty damn high before I'd put my money into some artificial currency," I say.

"Hell, they're all artificial," Joe says. "If you think about it, we've been doing this forever. We put our money in stocks, bonds, shares of mutual funds. Those things represent real assets–factories, ships, bananas, software, gold, whatever. Simoleons is just a new name for those assets. You carry around a smart card and spend it just like cash. Or else you go shopping in the Metaverse and spend the money online, and the goods show up on your doorstep the next morning."

I say, "Who's going to fall for that?"

"Everyone," he says. "For our big promo, we're going to give Simoleons away to some average Joes at the Super Bowl. We'll check in with them one, three, six months later, and people will see that this is a safe and stable place to put their money."

"It doesn't inspire much confidence," I say, "to hand the stuff out like Monopoly money."

He's ready for this one. "It's not a handout. It's a sweepstakes." And that's when he asks me to calculate how many jelly beans will fill Soldier Field.

Two hours later, I'm down at the local galaxy-class grocery store, in Bulk: a Manhattan of towering Lucite bins filled with steel-cut rolled oats, off-brand Froot Loops, sun-dried tomatoes, prefabricated s'mores, macadamias, French roasts and pignolias, all dispensed into your bag or bucket with a jerk at the handy Plexiglas guillotine. Not a human being in sight, just robot restocking machines trundling back and forth on a grid of over-head catwalks and surveillance cameras hidden in smoked-glass hemispheres. I stroll through the gleaming Lucite wonderland holding a perfect 6-in. cube improvised from duct tape and cardboard. I stagger through a glitter gulch of Gummi fauna, Boston baked beans, gobstoppers, Good & Plenty, Tart'n Tiny. Then, bingo: bulk jelly beans, premium grade. I put my cube under the spout and fill it.

Who guesses closest and earliest on the jelly beans wins the Simoleons. They've hired a Big Six accounting firm to make sure everything's done right. And since they can't actually fill the stadium with candy, I'm to come up with the Correct Answer and supply it to them and, just as important, to keep it secret.

I get home and count the beans: 3,101. Multiply by 8 to get the number in a cubic foot: 24,808. Now I just need the number of cubic feet in Soldier Field. My nephews are sprawled like pithed frogs before the HDTV, teaching themselves physics by lobbing antimatter bombs onto an offending civilization from high orbit. I prance over the black zigzags of the control cables and commandeer a unit.

Up on the screen, a cartoon elf or sprite or something pokes its head out from behind a window, then draws it back. No, I'm not a paranoid schizophrenic–this is the much-hyped intelligent agent who comes with the box. I ignore it, make my escape from Game-land and blunder into a lurid district of the Metaverse where thousands of infomercials run day and night, each in its own window. I watch an ad for Chinese folk medicines made from rare-animal parts, genetically engineered and grown in vats. Grizzly-bear gallbladders are shown growing like bunches of grapes in an amber fluid.

The animated sprite comes all the way out, and leans up against the edge of the infomercial window. "Hey!" it says, in a goofy, exuberant voice, "I'm Raster! Just speak my name–that's Raster–if you need any help."

I don't like Raster's looks. It's likely he was wandering the streets of Toontown and waving a sign saying WILL ANNOY GROWNUPS FOR FOOD until he was hired by the cable company. He begins flying around the screen, leaving a trail of glowing fairy dust that fades much too slowly for my taste.

"Give me the damn encyclopedia!" I shout. Hearing the dread word, my nephews erupt from the rug and flee.

So I look up Soldier Field. My old Analytic Geometry textbook, still flecked with insulation from the attic, has been sitting on my thigh like a lump of ice. By combining some formulas from it with the encyclopedia's stats…

"Hey! Raster!"

Raster is so glad to be wanted that he does figure eights around the screen. "Calculator!" I shout.

"No need, boss! Simply tell me your desired calculation, and I will do it in my head!"

So I have a most tedious conversation with Raster, in which I estimate the number of cubic feet in Soldier Field, rounded to the nearest foot. I ask Raster to multiply that by 24,808 and he shoots back; 537,824,167,717.

A nongeek wouldn't have thought twice. But I say, "Raster, you have Spam for brains. It should be an exact multiple of eight!" Evidently my brother's new box came with one of those defective chips that makes errors when the numbers get really big.

Raster slaps himself upside the head; loose screws and transistors tumble out of his ears. "Darn! Guess I'll have to have a talk with my programmer!" And then he freezes up for a minute.

My sister-in-law Anne darts into the room, hunched in a don't-mind-me posture, and looks around. She's terrified that I may have a date in here. "Who're you talking to?"

"This goofy I.A. that came with your box," I say. "Don't ever use it to do your taxes, by the way."

She cocks her head. "You know, just yesterday I asked it for help with a Schedule B, and it gave me a recipe for shellfish bisque."

"Good evening, sir. Good evening, ma'am. What were those numbers again?" Raster asks. Same voice, but different inflections–more human. I call out the numbers one more time and he comes back with 537,824,167,720.

"That sounds better," I mutter.

Anne is nonplussed. "Now its voice recognition seems to be working fine."

"I don't think so. I think my little math problem got forwarded to a real human being. When the conversation gets over the head of the built-in software, it calls for help, and a human steps in and takes over. He's watching us through the built-in videocam," I explain, pointing at the fish-eye lens built into the front panel of the set-top box, "and listening through the built-in mike."

Anne's getting that glazed look in her eyes; I grope for an analog analogy. "Remember The Exorcist? Well, Raster has just been possessed, like the chick in the flick. Except it's not just Beelzebub. It's a customer-service rep."

I've just walked blind into a trap that is yawningly obvious to Anne. "Maybe that's a job you should apply for!" she exclaims.

The other jaw of the trap closes faster than my teeth chomping down on my tongue: "I can take your application online right now!" says Raster.

My sister-in-law is the embodiment of sugary triumph until the next evening, when I have a good news/bad news conversation with her. Good: I'm now a Metaverse customer-service rep. Bad: I don't have a cubicle in some Edge City office complex. I telecommute from home from her home, from her sofa. I sit there all day long, munching through my dwindling stash of tax-deductible jelly beans, wearing an operator's headset, gripping the control unit, using it like a puppeteer's rig to control other people's Rasters on other people's screens, all over the U.S. I can see them–the wide-angle view from their set-top boxes is piped to a window on my screen. But they can't see me–just Raster, my avatar, my body in the Metaverse.

Ghastly in the mottled, flattening light of the Tube, people ask me inane questions about arithmetic. If they're asking for help with recipes, airplane schedules, child-rearing or home improvement, they've already been turfed to someone else. My expertise is pure math only.

Which is pretty sleepy until the next week, when my brother's agency announces the big Simoleons Sweepstakes. They've hired a knot-kneed fullback as their spokesman. Within minutes, requests for help from contestants start flooding in. Every Bears fan in Greater Chicago is trying to calculate the volume of Soldier Field. They're all doing it wrong; and even the ones who are doing it right are probably using the faulty chip in their set-top box. I'm in deep conflict-of-interest territory here, wanting to reach out with Raster's stubby, white-gloved, three-fingered hand and slap some sense into these people.

But I'm sworn to secrecy. Joe has hired me to do the calculations for the Metrodome, Three Rivers Stadium, RFK Stadium and every other N.F.L. venue. There's going to be a Simoleons winner in every city.

We are allowed to take 15-minute breaks every four hours. So I crank up the Home Theater, just to blow the carbon out of its cylinders, and zip down the main street of the Metaverse to a club that specializes in my kind of tunes. I'm still "wearing" my Raster uniform, but I don't care–I'm just one of thousands of Rasters running up and down the street on their breaks.

My club has a narrow entrance on a narrow alley off a narrow side street, far from the virtual malls and 3-D video-gram amusement parks that serve as the cash cows for the Metaverse's E-money economy. Inside, there's a few Rasters on break, but it's mostly people "wearing" more creative avatars. In the Metaverse, there's no part of your virtual body you can't pierce, brand or tattoo in an effort to look weirder than the next guy.

The live band onstage–jacked in from a studio in Prague–isn't very good, so I duck into the back room where there are virtual racks full of tapes you can sample, listening to a few seconds from each song. If you like it, you can download the whole album, with optional interactive liner notes, videos and sheet music.

I'm pawing through one of these racks when I sense another avatar, something big and shaggy, sidling up next to me. It mumbles something; I ignore it. A magisterial throat-clearing noise rumbles in the subwoofer, crackles in the surround speakers, punches through cleanly on the center channel above the screen. I turn and look: it's a heavy-set creature wearing a T shirt emblazoned with a logo HACKERS 1111. It has very long scythe-like claws, which it uses to grip a hot-pink cylinder. It's much better drawn than Raster; almost Disney-quality.

The sloth speaks: "537,824,167,720."

"Hey!" I shout. "Who the hell are you?" It lifts the pink cylinder to its lips and drinks. It's a can of Jolt. "Where'd you get that number?" I demand. "It's supposed to be a secret."

"The key is under the doormat," the sloth says, then turns around and walks out of the club.

My 15-minute break is over, so I have to ponder the meaning of this through the rest of my shift. Then, I drag myself up out of the couch, open the front door and peel up the doormat.

Sure enough, someone has stuck an envelope under there. Inside is a sheet of paper with a number on it, written in hexa-decimal notation, which is what computer people use: 0A56 7781 6BE2 2004 89FF 9001 C782–and so on for about five lines.

The sloth had told me that "the key is under the doormat," and I'm willing to bet many Simoleons that this number is an encryption key that will enable me to send and receive coded messages.

So I spend 10 minutes punching it into the set-top box. Raster shows up and starts to bother me: "Can I help you with anything?"

By the time I've punched in the 256th digit, I've become a little testy with Raster and said some rude things to him. I'm not proud of it. Then I hear something that's music to my ears: "I'm sorry, I didn't understand you," Raster chirps. "Please check your cable connections–I'm getting some noise on the line."

A second figure materializes on the screen, like a digital genie: it's the sloth again. "Who the hell are you?" I ask.

The sloth takes another slug of Jolt, stifles a belch and says, "I am Codex, the Crypto-Anarchist Sloth."

"Your equipment requires maintenance," Raster says. "Please contact the cable company."

"Your equipment is fine," Codex says. "I'm encrypting your back channel. To the cable company, it looks like noise. As you figured out, that number is your personal encryption key. No government or corporation on earth can eavesdrop on us now."

"Gosh, thanks," I say.

"You're welcome," Codex replies. "Now, let's get down to biz. We have something you want. You have something we want."

"How did you know the answer to the Soldier Field jelly-bean question?"

"We've got all 27," Codex says. And he rattles off the secret numbers for Candlestick Park, the Kingdome, the Meadowlands . . .

"Unless you've broken into the accounting firm's vault," I say, "there's only one way you could have those numbers. You've been eavesdropping on my little chats with Raster. You've tapped the line coming out of this set-top box, haven't you?"

"Oh, that's typical. I suppose you think we're a bunch of socially inept, acne-ridden, high-IQ teenage hackers who play sophomoric pranks on the Establishment."

"The thought had crossed my mind," I say. But the fact that the cartoon sloth can give me such a realistic withering look, as he is doing now, suggests a much higher level of technical sophistication. Raster only has six facial expressions and one of them is very good.

"Your brother runs an ad agency, no?"

"Correct."

"He recently signed up Simoleons Corp.?"

"Correct."

"As soon as he did, the government put your house under full-time surveillance."

Suddenly the glass eyeball in the front of the set-top box is looking very big and beady to me. "They tapped our infotainment cable?"

"Didn't have to. The cable people are happy to do all the dirty work–after all, they're beholden to the government for their monopoly. So all those calculations you did using Raster were piped straight to the cable company and from there to the government. We've got a mole in the government who cc'd us everything through an anonymous remailer in Jyväskylä, Finland."

"Why should the government care?"

"They care big-time," Codex says. "They're going to destroy Simoleons. And they're going to step all over your family in the process."

"Why?"

"Because if they don't destroy E-money," Codex says, "E-money will destroy them."

The next afternoon I show up at my brother's office, in a groovily refurbished ex-power plant on the near West Side. He finishes rolling some calls and then waves me into his office, a cavernous space with a giant steam turbine as a conversation piece. I think it's supposed to be an irony thing.

"Aren't you supposed to be cruising the I-way for stalled motorists?" he says.

"Spare me the fraternal heckling," I say. "We crypto-anarchists don't have time for such things."

"Crypto-anarchists?"

"The word panarchist is also frequently used."

"Cute," he says, rolling the word around in his head. He's already working up a mental ad campaign for it.

"You're looking flushed and satisfied this afternoon," I say. "Must have been those two imperial pints of Hog City Porter you had with your baby-back ribs at Divane's Lakeview Grill."

Suddenly he sits up straight and gets an edgy look about him, as if a practical joke is in progress, and he's determined not to play the fool.

"So how'd you know what I had for lunch?"

"Same way I know you've been cheating on your taxes."

"What!?"

"Last year you put a new tax-deductible sofa in your home office. But that sofa is a hide-a-bed model, which is a no-no."

"Hackers," he says. "Your buddies hacked into my records, didn't they?"

"You win the Stratolounger."

"I thought they had safeguards on these things now."

"The files are harder to break into. But every time information gets sent across the wires—like, when Anne uses Raster to do the taxes—it can be captured and decrypted. Because, my brother, you bought the default data-security agreement with your box, and the default agreement sucks."

"So what are you getting at?"

"For that," I say, "we'll have to go someplace that isn't under surveillance."

"Surveillance!?" What the . . ." he begins. But then I nod at the TV in the corner of his office, with its beady glass eye staring out at us from the set-top box.

We end up walking along the lakeshore, which, in Chicago in January, is madness. But we hail from North Dakota, and we have all the cold-weather gear it takes to do this. I tell him about Raster and the cable company.

"Oh, Jesus!" he says. "You mean those numbers aren't secret?"

"Not even close. They've been put in the hands of 27 stooges hired by the government. The stooges have already FedEx'd their entry forms with the correct numbers. So, as of now, all of your Simoleons–$27 million worth–are going straight into the hands of the stooges on Super Bowl Sunday. And they will turn out to be your worst public-relations nightmare. They will cash in their Simoleons for comic books and baseball cards and claim it's safer. They will intentionally go bankrupt and blame it on you. They will show up in twos and threes on tawdry talk shows to report mysterious disappearances of their Simoleons during Metaverse transactions. They will, in short, destroy the image–and the business–of your client. The result: victory for the government, which hates and fears private currencies. And bankruptcy for you, and for Mom and Dad."

"How do you figure?"

"Your agency is responsible for screwing up this sweepstakes. Soon as the debacle hits, your stock plummets. Mom and Dad lose millions in paper profits they've never had a chance to enjoy. Then your big shareholders will sue your ass, my brother, and you will lose. You gambled the value of the company on the faulty data-security built into your set-top box, and you as a corporate officer are personally responsible for the losses."

At this point, big brother Joe feels the need to slam himself down on a park bench, which must feel roughly like sitting on a block of dry ice. But he doesn't care. He's beyond physical pain. I sort of expected to feel triumphant at this point, but I don't.

So I let him off the hook. "I just came from your accounting firm," I say. "I told them I had discovered an error in my calculations–that my set-top box had a faulty chip. I supplied them with 27 new numbers, which I worked out by hand, with pencil and paper, in a conference room in their offices, far from the prying eye of the cable company. I personally sealed them in an envelope and placed them in their vault."

"So the sweepstakes will come off as planned," he exhales. "Thank God!"

"Yeah–and while you're at it, thank me and the panarchists," I shoot back. "I also called Mom and Dad, and told them that they should sell their stock–just in case the government finds some new way to sabotage your contest."

"That's probably wise," he says sourly, "but they're going to get hammered on taxes. They'll lose 40% of their net worth to the government, just like that."

"No, they won't," I say. "They aren't paying any taxes."

"Say what?" He lifts his chin off his mittens for the first time in a while, reinvigorated by the chance to tell me how wrong I am. "Their cash basis is only $10,000–you think the IRS won't notice $20 million in capital gains?"

"We didn't invite the IRS," I tell him. "It's none of the IRS's damn business."

"They have ways to make it their business."

"Not any more. Mom and Dad aren't selling their stock for dollars, Joe."

"Simoleons? It's the same deal with Simoleons–everything gets reported to the government."

"Forget Simoleons. Think CryptoCredits."

"CryptoCredits? What the hell is a CryptoCredit?" He stands up and starts pacing back and forth. Now he's convinced I've traded the family cow for a handful of magic beans.

"It's what Simoleons ought to be: E-money that is totally private from the eyes of government."

"How do you know? Isn't any code crackable?"

"Any kind of E-money consists of numbers moving around on wires," I say. "If you know how to keep your numbers secret, your currency is safe. If you don't, it's not. Keeping numbers secret is a problem of cryptography–a branch of mathematics. Well, Joe, the crypto-anarchists showed me their math. And it's good math. It's better than the math the government uses. Better than Simoleons' math too. No one can mess with CryptoCredits."

He heaves a big sigh. "O.K., O.K.–you want me to say it? I'll say it. You were right. I was wrong. You studied the right thing in college after all."

"I'm not worthless scum?"

"Not worthless scum. So. What do these crypto-anarchists want, anyway?"

For some reason I can't lie to my parents, but Joe's easy. "Nothing," I say. "They just wanted to do us a favor, as a way of gaining some goodwill with us."

"And furthering the righteous cause of World Panarchy?"

"Something like that."

Which brings us to Super Bowl Sunday. We are sitting in a skybox high up in the Superdome, complete with wet bar, kitchen, waiters and big TV screens to watch the instant replays of what we've just seen with our own naked, pitiful, nondigital eyes.

The corporate officers of Simoleons are there. I start sounding them out on their cryptographic protocols, and it becomes clear that these people can't calculate their gas mileage without consulting Raster, much less navigate the subtle and dangerous currents of cutting-edge cryptography.

A Superdome security man comes in, looking uneasy. "Some, uh, gentlemen here," he says. "They have tickets that appear to be authentic."

It's three guys. The first one is a 300 pounder with hair down to his waist and a beard down to his navel. He must be a Bears fan because he has painted his face and bare torso blue and orange. The second one isn't quite as introverted as the first, and the third isn't quite the button-down conformist the other two are. Mr. Big is carrying an old milk crate. What's inside must be heavy, because it looks like it's about to pull his arms out of their sockets.

"Mr. and Mrs. De Groot?" he says, as he staggers into the room. Heads turn towards my mom and dad, who, alarmed by the appearance of these three, have declined to identify themselves. The guy makes for them and slams the crate down in front of dad.

"I'm the guy you've known as Codex," he says. "Thanks for naming us as your broker."

If Joe wasn't a rowing-machine abuser, he'd be blowing aneurysms in both hemispheres about now. "Your broker is a half-naked blue-and-orange crypto-anarchist?"

Dad devotes 30 seconds or so to lighting his pipe. Down on the field, the two-minute warning sounds. Dad puffs out a cloud of smoke and says, "He seemed like an honest sloth."

"Just in case," Mom says, "we sold half the stock through our broker in Bismarck. He says we'll have to pay taxes on that."

"We transferred the other half offshore, to Mr. Codex here," Dad says, "and he converted it into the local currency–tax free."

"Offshore? Where? The Bahamas?" Joe asks.

"The First Distributed Republic," says the big panarchist. "It's a virtual nation-state. I'm the Minister of Data Security. Our official currency is CryptoCredits."

"What the hell good is that?" Joe asks.

"That was my concern too," Dad says, "so, just as an experiment, I used my CryptoCredits to buy something a little more tangible."

Dad reaches into the milk crate and heaves out a rectangular object made of yellow metal. Mom hauls out another one. She and Dad begin lining them up on the counter, like King and Queen Midas unloading a carton of Twinkies.

It takes Joe a few seconds to realize what's happening. He picks up one of the gold bars and gapes at it. The Simoleons execs crowd around and inspect the booty.

"Now you see why the government wants to stamp us out," the big guy says. "We can do what they do–cheaper and better."

For the first time, light dawns on the face of the Simoleons CEO. "Wait a sec," he says, and puts his hands to his temples. "You can rig it so that people who use E-money don't have to pay taxes to any government? Ever?"

"You got it," the big panarchist says. The horn sounds announcing the end of the first half.

"I have to go down and give away some Simoleons," the CEO says, "but after that, you and I need to have a talk."

The CEO goes down in the elevator with my brother, carrying a box of 27 smart cards, each of which is loaded up with secret numbers that makes it worth a million Simoleons. I go over and look out the skybox window: 27 Americans are congregated down on the 50-yard line, waiting for their mathematical manna to descend from heaven. They are just the demographic cross section that my brother was hoping for. You'd never guess they were all secretly citizens of the First Distributed Republic.

The crypto-anarchists grab some Jolt from the wet bar and troop out, so now it's just me, Mom and Dad in the skybox. Dad points at the field with the stem of his pipe. "Those 27 folks down there," he says. "They didn't get any help from you, did they?"

I've lied about this successfully to Joe. But I know it won't work with Mom and Dad. "Let's put it this way," I say, "not all panarchists are long-haired, Jolt-slurping maniacs. Some of them look like you–exactly like you, as a matter of fact."

Dad nods; I've got him on that one.

"Codex and his people saved the contest, and our family, from disaster. But there was a quid pro quo."

"Usually is," Dad says.

"But it's good for everyone. What Joe wants–and what his client wants–is for the promotion to go well, so that a year from now, everyone who's watching this broadcast today will have a high opinion of the safety and stability of Simoleons. Right?"

"Right."

"If you give the Simoleons away at random, you're rolling the dice. But if you give them to people who are secretly panarchists—who have a vested interest in showing that E-money works it's a much safer bet."

"Does the First Distributed Republic have a flag?" Mom asks, out of left field. I tell her these guys look like sewing enthusiasts. So, even before the second half starts, she's sketched out a flag on the back of her program. "It'll be very colorful," she says. "Like a jar of jelly beans."

©1995 by Neal Stephenson. First appeared in the Spring 1995 issue of *Time*.

1 SHIFTING PARADIGMS

Cash. Moola. Boodle. Simoleons. Smackers. Wampum. Bucks. Bread. Dough. Lettuce. Kale. Green. Whatever it's called, it has become the grease that keeps the machinery of society from grinding to a standstill. Whatever it's called, it's always been something tactile, something to be felt, dealt, touched, handled—like the goods in the barter system it supplanted. There's always been a measure of security in that, a security needed by humanity.

The barter system was straightforward. I'll trade you this dead sheep for your handful of corn seeds. Both sides got to see, touch, and smell the end result of their trade. Obviously, barter has limitations. Try dragging an elk through Union Square to buy a silk tie and it becomes obvious that this technique isn't useful for most transactions.

Worse, it's too easy to get ripped off during a barter deal. You don't know if the animal is rancid, the seeds irradiated, or the car engine rebuilt. And it's virtually impossible to throw services into the mix. In the face of these difficulties, the Romans (or maybe it was the Greeks) invented cash.

Cash is just a placeholder. It's a mass illusion, a way for an entire society to agree on the value of a worthless icon and maintain trade by continuing to value the icon. If you've ever returned from a trip with a bunch of kopecks or rubles in your pocket, it quickly becomes clear that money is worth only as much as the person accepting it says it's worth.

Cash is a paradigm. A set of rules, assumptions, and patterns that society follows to make it easier for all of us. There are societies where the paradigm of cash is not accepted, but generally speaking, it's been so useful that people have put their faith into the little pieces of paper, going so far as to refer to it as "almighty."

Once cash was invented, a thousand variations on the concept became possible. Banking, for example. Unlike barter, cash created a store of value. A way to increase your assets without a warehouse. With some people accumulating cash, and others wanting to borrow it, the idea of interest arose. So not only did people make a leap of faith and decide that an almost worthless piece of metal or sheet of paper was exchangeable for goods and services, they went further and said, "If you give me your cash, I promise that I'll give you back that same cash, plus more, some time in the future."

With the paradigm of interest in place, whole industries can spring up. A business can grow faster than its assets allow, because it can borrow money. Investment banks can act as middlemen. Farmers can plan for bad crops. And so on. Virtually every facet of our society is affected by the existence of cash, and is possible because of the existence of interest. But if you look at it closely, you'll realize that it's all a fiction, that the emperor is wearing no clothes.

As long as we continue to accept the paradigm of cash, society will continue to function. But a new development is going to change that paradigm forever, and it offers a fundamental change to just about every element of our economy.

The new paradigm will be, for some at least, as big a leap as the acceptance of cash was for the Greeks (or the Romans). Now we're being asked to make another leap of faith when it comes to what's exchanged for goods and services, a jump so great that comparing it to the leap from barter to cash would be like comparing springing over the Grand Canyon to stepping over a crack in the sidewalk. Digital cash requires us to eliminate the last element of barter that remains in our system—the tactile

Shifting Paradigms

pleasure of holding it. Cash is about to become invisible, fungible, transferable, and almost magic.

These changes, which signal the dawn of the Information Age and a shift away from the industrial paradigm, aren't limited to how companies do business and sell what they make. They go much deeper than that, right to the roots of society itself and to its sap, money. Joel Kurtzman, a former executive editor of the *Harvard Business Review*, put it this way in his insightful book *The Death of Money*: "Money has been transmogrified. It is no longer a thing, an object you can dig up at the beach or search for behind the cushions of a sofa; it is a system. Money is a network that comprises hundreds of thousands of computers of every type, wired together in places as lofty as the Federal Reserve—which settles accounts between banks every night that are worth trillions of dollars—and as mundane as the thousands of gas pumps around the world outfitted to take credit and debit cards.

"In the new world of money, even the largest banks no longer need vaults," Kurtzman notes. "Instead, they store their money on disk drives and computer tapes, and they protect those funds not by hiring brawny guards but by employing brainy Ph.D. mathematicians and software specialists to write secret codes."

At first, digital cash seems like a logical evolution. But a close look will demonstrate that it is one of the most fundamental paradigm shifts we will see in our lifetime.

How fundamental? Consider some past paradigm shifts:

The invention of the internal combustion engine led to the creation of the car. The car eliminated obvious things (buggy-whip makers, carriage men, stables, etc.) and created a whole host of new industries (road construction, service stations, tire recycling, auto-glass makers, etc.) It also created ripple effects that we're still feeling. Without the car, Ray Kroc couldn't have built McDonald's (who needs fast food if it takes an hour to get there to pick it up?) nor would there be large rock concerts and sporting events (where do you keep 80,000 horses?)

Before the introduction of the internal combustion engine, distances were measured in days, not minutes. It took weeks to travel to another country, an afternoon to buy a loaf of bread. But there were less obvious changes that came with the introduction of the car. Highways needed to be built (contractors profited). Real estate values were turned upside down (real estate agents and savvy investors profited). Fast food, supermarkets, movie theaters, suburbs, large sporting events, and NASCAR racing all became possible as well. If you could have foreseen

> **"Money has been transmogrified. It is no longer a thing, an object you can dig up at the beach or search for behind the cushions of a sofa; it is a system."**
> Joel Kurtzman

the successful introduction of the car, how would you have responded? That's the key to taking advantage of a paradigm shift.

The shores of Lake Erie are littered with the hulks of large industrial companies that ignored the last paradigm shift and weren't quick enough to respond when it finally reached them.

It's easy to argue that the invention of the car shook our societal paradigms to their roots. But simpler inventions like the fax and FedEx have had similar impact. These rapid communication tools did more than make it easy to get a contract signed. They fundamentally changed the way business is done by companies large and small. They introduced speed as a very real competitive element. Companies that couldn't respond withered, while those that could take advantage of rapid information exchange and rapid turnaround have prospered.

Because paradigm shifts are so powerful, there are a lot of people crying wolf. Three years ago, every journalist and Hollywood executive was busy crowing about interactive TV. Before that, it was videophones, personal airplanes, and transportation by blimp.

So how do you tell the difference between the would-be shifts and the real thing? Hard to say. One clue is the speed of adoption. The microwave oven was a shift waiting to happen, and it took just a few years for it to gain speed. Same as the explosive growth of cable TV. When society and technology are ready at the same time, paradigm adoption can be quite rapid.

This book is your handbook to the key element of the most important paradigm shift for the near future. The way we buy things (and what we buy) is about to change forever. Speed, information, and loyalty will become infinitely more important, while traditional factors like location, capital, and workforce size will become almost trivial.

The move to digital cash is already happening. Best-selling business books talk about the one-to-one relationship with consumers, guerrilla marketing and bypassing traditional media, reengineering the corporation and taking advantage of new technology. We blithely use our credit card for a ninety-nine-cent bag of potato chips, forgetting for the moment that the system that allows us to do so cost billions of dollars to create and would have been impossible just a decade ago.

Businesses have discovered just how powerful speed is. If a company can cut the development time of a new product in half, from two years to one, it can turn out ten generations of product improvements in each decade, not just five.

> **It's easy to argue that the invention of the car shook our societal paradigms to their roots. But simpler inventions like the fax and FedEx have had similar impact.**

The Japanese were the first to embrace speed. In 1965, a Toyota was a primitive, Third World car. The Japanese were at least thirty years behind Detroit in their engineering and development. They chose to compete by cutting the fat from the development cycle, turning out cars far quicker than the United States could.

Obviously, the strategy worked—and soon spread to dozens of other industries. The goal: to eliminate slack from the system and improve quality and technology as quickly as possible.

Prescription eyeglass lenses are ground and placed in custom frames within sixty minutes by companies like Lenscrafters and Pearle Vision Express. Polaroid gives us sixty-second prints, instant photo shops offer high-quality 35-millimeter prints in sixty minutes. Camcorders make instant home movies. Personal computers and laser printers produce instant printing. These instant products go hand in hand with instant services: the ten-minute oil change, instant cash from an ATM, travel reservations in seconds, fast-food tacos in less than twenty seconds. "What these products and services have in common is that they deliver instant customer gratification in a cost-effective way," William Davidow and Michael Malone write in *The Virtual Corporation*.

The speed revolution has almost engulfed the consumer. Products are now delivered by Federal Express (you can even get your L. L. Bean socks delivered by Christmas if you order by the twenty-fourth!) and credit cards have become ubiquitous. But the speed connection to the consumer has been a morass. A package from New York to Rochester goes through Memphis. A credit card number is handled by dozens of people, and then its owner can be the victim of fraud or a chargeback.

In order to make the connection between creators and consumers as swift and efficient as possible, we need an instant, certain, auditable way to deliver and pay for goods. Many products and services are quickly becoming digital, allowing distribution via computer. This book is about the second half of the equation: instant payment for these products.

Developing in parallel to these virtual organizations are "virtual products." According to Davidow and Malone, these products already exist in our daily lives. Businesses exist that have no inventory and few full-time employees.

At the same time, companies are realizing the benefits of outsourcing labor and production. Technology no longer requires knowledge workers to be onsite—data can be sent anywhere. Rapid delivery systems no longer require all parts of a factory to be under one roof—Porsche now buys completed dashboards from a distant supplier instead of painstakingly assembling them onsite.

> **This book is your handbook to the key element of the most important paradigm shift for the near future. The way we buy things (and what we buy) is about to change forever.**

Outsourcing labor creates two interesting side effects. First, the company becomes more wedded to its supplier, because it's so dependent, but at the same time, they become pickier and more likely to switch to a better alternative, since the cost of switching is small compared to the cost of closing down an office or factory.

Suddenly, it's possible to build a company that acts like a $20 million operation but only has one or two "employees." ABB runs an organization with 33,000 people worldwide with a home office of just forty-two people. Insurance companies send their completed applications to the Philippines for inputting, and publishers publish books by authors they've never met or even spoken to on the telephone.

A big company with a full-time in-house staff of designers, artists, fabricators, or engineers is always going to have trouble keeping up with an alert competitor with a big fat Rolodex. The freelancers can create better products, faster, because the competitor can choose exactly the right people to staff a team. Not only does the speed mean greater sales, it actually costs less than the old-fashioned way.

The increase in speed can also lead to an increase in quality. And the quality can actually save the company money. Because an offsite contributor is responsible for the quality of the component being delivered (avoiding the "we'll fix it later" syndrome) inspection is built in to every step of the process, and the final product is bound to work better.

So the challenge to companies is to harness technology that l quality (quality is free) and speed (speed is free) and turn these ' ̠e" assets into competitive weapons that will demolish slower competitors. That leads to the many changes this paradigm is bringing us: EDI, FedEx, computer-controlled manufacturing, teams, suppliers with offices in customer's offices, and all the other techniques you're reading about.

There have been dozens of books that outline the groundswell that's leading to this powerful paradigm shift. The important message here is, THIS AIN'T JUST A TECHNICAL GIMMICK, GUYS! Computers are facilitating the process, but it isn't happening because of computers.

Society is ready for this shift. Businesses are itching to take advantage of the velocity, flexibility, and high customer involvement that the new era will bring. Everything is in place except for one key element: There's no easy way to pay for all this stuff. Credit cards weren't made to do it, and they're doing it badly. Once digital cash is in place, once the circle is completed, expect an explosion in the speed of adoption, and a similar explosion in the profits recorded by those smart enough to get in early.

2 THE INTERNET

This section covers elements of the Internet that are important to understanding digital cash and the way it will work. If you know your way around the Net, feel free to jump to the next chapter. The most important things to understand about the Internet are:

- It's very, very fast.

- The more it grows, the faster it can get.

- Nobody owns it.

- It's not secure, and wasn't meant to be.

- There are a bunch of acronyms, and it helps to know them

- E-mail is the underlying engine of the whole shebang. Everything else is just enhancement.

The Internet in Four Paragraphs

Computers used to work in isolation. They were huge, expensive machines, tended by acolytes and fed punch cards. Eventually, it became efficient to wire several computers to each other, allowing them to exchange information. When two or more computers are connected, the result is called a network.

Most computers, especially mainframes, spend most of their time idle. They rarely operate at peak performance on a regular basis. Computer scientists struggled to find something for the computer to do with this idle time.

A network of networked computers is ideal. Computers can now exchange information with dozens or thousands of other computers, pitching in to help as they have spare time.

This network of networks was originally designed to allow scientists to access expensive supercomputers. But it has evolved into a system that allows more than 35 million people to communicate with each other in a myriad of ways.

The Evolution of the Internet

The seeds of the Internet were planted about twenty years ago in something called ARPANET. ARPANET was a U.S. Defense Department network designed for military research. On ARPANET, all computers were created equal, not because its designers were democratically minded but because they assumed that in the event of a catastrophe, network connections would be unreliable. Hence they wanted the computers to be peers who could talk to each other regardless of what happened to the network. To this end, communications overhead, which usually resides in the network itself, was placed in each computer client. And communications between clients could be accomplished with a minimum of information. A computer need only put the data it wished to send through the network in an "envelope," called an Internet protocol (IP) packet, and "address" the packets properly.

This technique of exchanging packets of data, called packet switching, was an innovative departure from other schemes used at the time, and it is still used by communications networks today. In a packet-switching system, data are broken into small chunks of information, given an address of origination and a destination address, and sent on

> **ARPANET was a U.S. Defense Department network designed for military research. On ARPANET, all computers were created equal, not because its designers were democratically-minded but because they assumed that in the event of a catastrophe, network connections would be unreliable.**

their way down the network. As the packets arrive at their destination, they're reassembled. If anything happens to a packet in its travels—if it's lost or corrupted—the destination computer will ask the originating computer to resend the packet.

Internet developers began installing IP software on every kind of computer platform. This worked well because many of the organizations connecting to ARPANET didn't have uniform policies about computer purchasing. Departments, offices, bureaus, and individuals bought any box they felt like buying and expected it to work on the Net. Sure enough, IP became the only practical way for computers from different manufacturers to communicate.

Since ARPANET was a military network, its designers had to plan for the worst. They had to take into account that the network might be called on to operate under the most unreliable conditions. To cope with this contingency, they created something called the Host-to-Host Protocol. The problem with this protocol was that it severely restricted the number of computers that could be on the network. So in 1972 the designers began working on the next generation of protocols. The result was the Transmission Control Protocol/Internet Protocol, or TCP/IP, a collection of more than a hundred protocols which is still used today. TCP/IP became an official Internet standard in 1983, although it had been being used on a voluntary basis since 1979.

While TCP/IP was being developed, Ethernet local area networks (LANs) began to emerge. Ethernet, which is probably used to connect your computer to your coworkers', created efficient networks within organizations. This created new pressure within data-processing departments to give these LANs direct access to ARPANET without being at the mercy of a single large time-sharing computer.

As these LANs were gaining access to ARPANET, other organizations, like the National Science Foundation (NSF), began building their own networks using IP or its close relatives. In the late '80s, NSFNET, in a bold move, established five regional supercomputer centers. Up to that time, only weapons developers and a few researchers at large corporations had access to the world's fastest computers. With the establishment of the NSF centers, these computing resources would be available for any scholarly research. The foundation also promoted universal educational access by agreeing to fund campus connections to the network only if the school had a plan to spread access around. The NSF had to limit the number of centers it could set up because they were so expensive. This meant the foundation had to find a way to network the facilities. It tried to work with ARPANET, but the red tape was overwhelming. The NSF threw up its hands and decided to build its own network.

The foundation adopted ARPANET's IP technology and connected the centers using phone lines that could move data at 56,000 bits per second (bps), or about two typewritten pages a second. That created another snag. Connecting every campus in the country to its closest center with 56K bps phone lines would cost a fortune. So the foundation created a daisy-chain arrangement. Each school in a region was connected to a neighboring school. Then the entire chain was connected to its regional computing center at a single point. The scheme worked fine for a while, but by the late '80s, traffic on the system brought the computers managing the network and the phone lines tying it together to their knees. The system was upgraded in 1987 by Merit Network, which ran Michigan's educational network in partnership with IBM and MCI. During the upgrade, phone-line speeds were improved by a factor of 20 and faster computers were installed to control the network.

By the end of the 1980s, the basic framework of the Internet was set. Here was a robust system, capable of growth. As new computers came online, each helped to share the load, actually making the network more efficient.

Just as important as the infrastructure was the culture. Because of its nonprofit origins, the Net has a tradition of helpfulness and community. As new people are added, many are quickly indoctrinated to the mores of the society. Just as the Grateful Dead attracted fans who all looked as if they were somehow related, the culture of the Net has led to a surprisingly homogeneous society. This society eases the occasional snafus caused by traffic and computer glitches, but it also masks some of the challenges necessary to create a truly working system—one that doesn't depend on the kindness of strangers.

Problems with traffic overload continue today and will continue into the future, even though there is a greater awareness of the importance of creating a state-of-the-art information superstructure. With most four-year colleges and universities connected to the Net, there's pressure to connect the rest of the education system—primary and secondary schools.

Given the huge quantities of free data available online, experts project that most school districts will have some sort of online access by the end of the century. With access arriving at most schools, the entire

> The culture of the Net has led to a surprisingly homogeneous society. This society eases the occasional snafus caused by traffic and computer glitches, but it also masks some of the challenges necessary to create a truly working system—one that doesn't depend on the kindness of strangers.

infrastructure by which educational material is delivered can change, becoming much cheaper and more efficient.

Moreover, graduates who are familiar with the Internet from their college days are persuading their employers to get wired. Meanwhile, non-IP networks are getting into the act. Initially these oddball nets were connected to the Internet via gateways and their capabilities were pretty much limited to the exchange of electronic mail. As the cost of access goes down, expect that most users will upgrade their speed and connectivity.

George Gilder has proclaimed that "bandwidth will be free," and he's probably correct. While the speed of phone-based modems has probably reached a theoretical limit, look for cable TV or wireless modems to give every household, every business, every school high-speed access to the Net.

This phenomenal growth of the Internet, however, will be lost to the public pioneer networks that began it all. In April 1995, the backbone of the last of those networks, NSFNET, was retired. A commercial system of backbones has replaced what was once a totally government-sponsored system. The cornerstones of the Internet—ARPANET, SATNET, PRNET, and NSFTNET—are gone, but what they wrought continues to flourish.

The Internet Today

According to the Internet Society, the Internet today is a web of public and private networks sprawling over 50,000 networks in ninety countries, with electronic mail gateways extending the net's reach to 160 countries. By the end of 1994, some 5 million computers and estimated 20 million to 40 million users were reachable through the Internet. Network growth has been estimated at 10 percent per month.

The impact of the Internet extends beyond the almost numbing quantity of mentions in the media. It has profoundly changed the way individuals communicate with each other. It threatens to become the future of television (interactive TV delivered by living room CPU) as well as the future of the telephone (free long-distance service).

There are currently more than 10,000 companies with pages on the World Wide Web, and the number of domains registered increases every day. There have even been several hotly contested lawsuits over the use of brand names on the Net.

Despite the huge attention being paid to the Net, there's no chance that it will become a bureaucracy or a public company. There's no president, chief operating officer, or Net czar. The Net's direction is determined by the Internet Society, or ISOC, a voluntary membership organization pledged to promote global information exchange through Internet technology. Basically, though, the Net is rulerless. There's no one in charge. Sort of like Dodge City before the sheriff came to town.

This unique environment creates a fascinating test bed for digital cash. Basically, the Net is

- self policing

- free

There's no barrier to get in, no government control, no barrier to get out. This really is the Wild West.

Is the Net really self-policing? Consider the case of the Green Card Lawyers, two marketers from Arizona who sent uninvited messages to hundreds of thousands of people on the Net. In return, they were flamed more than 30,000 times. Thirty thousand people took the time to send angry mail complaining about this abuse of the system.

This self-policing mechanism is surprisingly strong. When more than a million America Online members were set free on the Net in just one week, the community banded together to quickly teach the newbies the ropes. Primitive perhaps, certainly reminiscent of a small town, but it worked.

Accessing the Internet

Access to the Internet is usually achieved in one of four ways:

- Host access—a user makes a direct connection to the Internet; the user's computer becomes part of the Internet.

- Terminal access—a user connects to a host that is connected to the Internet.

- Network access—a network of computers is directly connected to the Internet.

- Gateway access—a non-Internet network, online service, or e-mail service connects to the Internet through a "gateway," which makes the information coming from those services compatible with the requirements of the Internet.

For some forms of digital cash, it doesn't matter which of the four methods the user decides on. By relying on e-mail, the lowest common denominator of Internet connectivity, these systems are universal. Other protocols require a real-time connection between the consumer and the merchant. Information—such as a conversation—is exchanged in order to verify identity and maintain security. While these systems may be more robust, they are by their nature more expensive to install (each consumer requires some sort of software or hardware device) and lay no claim to universality.

Many users connect to the Internet through a dialup connection. To make a dialup connection to the Internet, a modem is needed. Most modems in use today are significantly faster than those used four years ago, but remain quite slow compared to industrial-strength connections. Once cable TV lines are able to carry these signals, though, the speeds will increase dramatically.

The alternative to using conventional phone lines is using digital ones. Data transfer speeds using digital lines can exceed speeds on conventional lines by at least a factor of 2. But digital lines are very expensive, although the price of all-digital service in the form of ISDN has been falling.

Here's a quick overview of the basic tasks that can be accomplished once you're online.

Electronic Mail

You can strike up a correspondence with anyone who has an Internet address. This includes millions of users of online systems such as Prodigy and America Online. Electronic mail is infinitely faster than mail delivered by conventional post. Because the e-mail can be sent and read by a computer, it is far more flexible than traditional mail.

For example, a user can set her e-mail box to look for a letter from Jack and instantly respond with a pre-written message. She can send her Christmas announcement letter to fifty friends simultaneously. She can set mail to forward to another box.

Marketers can get even more out of e-mail. Customized mass mailings can be sent overnight at no cost. Robots can be set up to search for data near and far and report back. Everything from school closings to homework assignments to work product flow to government forms to your taxes can be sent by e-mail.

Internet mailing lists allow people with similar interests to exchange information and opinions about those interests. When you subscribe to a list, by sending a message to a program called a list server on a remote computer, you receive all the mail sent to the list by its subscribers. Lists can be on any subject from repetitive stress injuries to personal finance.

E-mail offers some downsides. It's not secure. A clever (or even a not-clever but persistent) hacker can read most of the mail you send and receive. Fortunately, this doesn't happen often, since the vast majority of mail isn't worth reading.

E-mail is also automatic. That means that overeager marketers can send too much mail to too many people. Individuals can counter this by creating filters that eliminate mail from unknown sources, but of course this can cripple the very access that makes e-mail useful.

Telnet

This service allows you to sit at one computer and log on to another computer elsewhere on the Internet. The computer may be on the other side of a campus or the other side of the world. When you're connected to the other computer through Telnet, you can use any of its services by issuing commands through your keyboard. So, for example, you could log into the card catalog of a university library miles away and do a topic search.

Telnet is a real-time connection. You are at the other computer. The connection puts far more stress on the Net. This connection is the basis of the World Wide Web (WWW), though, and of real-time phone conversation, which happens as we speak. Understanding the critical differences between Telnet and e-mail is crucial—Telnet is the live connection that's somewhat less reliable but far richer.

FTP

This stands for File Transfer Protocol. It's used to download files from a remote computer to your computer. FTP is machine-independent. It doesn't matter if the computers engaged in the transfer are the same brand or have the same operating system or anything else. If they can "talk" FTP, they can "walk" files between them.

FTP is a powerful tool for software distribution. Netscape, the now legendary start-up, was able to distribute millions of copies of its free software by FTP. This enabled them to establish a standard at virtually no cost.

> **Netscape, the now legendary start-up, was able to distribute millions of copies of its free software by FTP. This enabled them to usurp a standard at virtually no cost.**

Usenet

This is one of the most popular features of the Internet. There are over five thousand Usenet groups on the Internet. Like mailing lists, Usenet groups are discussion groups on a variety of subjects. Unlike mailing lists, which deliver the content of their discussions to your electronic mailbox every day, Usenet requires that you navigate to it and view the latest messages in your favorite groups online.

In an active newsgroup, there can be hundreds of messages filed every day. A good newsreader—the software used to read newsgroups' messages—will sort a group's messages into "threads." A thread is a topic, defined by the subject line in a message, within a newsgroup. A newsreader will sort all the messages for a given thread and store them with that thread. If a user isn't interested in a thread, he can ignore it and all the messages that go with it.

Usenet groups are a valuable self-selection mechanism, much treasured by marketers. They allow the anonymous world of the Net to fracture into smaller villages, where like-minded individuals can meet and discuss issues of mutual interest.

Gopher

This is a sort of universal front-end for the Internet. It was developed in 1991 at the University of Minnesota (hence the name "Gopher," the mascot of the college) but became so popular, it quickly spread throughout the Internet. It is a menu-based system that makes accessing the Internet's resources easier.

Gopher can establish connections with resources throughout the Internet. It does so automatically and while maintaining a consistent look. Its hierarchical menus are easy to negotiate with a computer's cursor keys or a mouse and some one-letter commands. When you choose a resource that would normally take several steps to access—Telnet or FTP, for example—Gopher automatically performs the relevant commands for you. You don't have to worry about things like domain names, IP addresses, or changing programs.

WAIS

As you might imagine, sometimes searching the Internet for file names or brief file descriptions won't yield much fruitful information. Or worse, it *will* yield an avalanche of irrelevant data. WAIS (pronounced "ways"), or Wide Area Information Server, addresses this problem. Every word in a document on a WAIS server is indexed. That allows a user to search for a word or string of words in those documents. That's the good news. The bad news is that WAIS is still in its infancy and only a small amount of the information on the Internet is indexed under the scheme.

World Wide Web

The hottest sector of the Internet is the World Wide Web. (A recent poll of 522 readers by *Network World*, a weekly computer trade newspaper, showed 40 percent of the respondents had Web sites in place and 29 percent said they were building a Web site or planned to build one.) It's so hot that it has been called the "killer ap"—an application so powerful that it drives an entire technology—of the Internet. There are several reasons why the Web has fired enthusiasm about cyberspace.

It looks better. While nothing on the Internet looks good, the WWW is certainly a step in the right direction. Until the Web made its debut, interacting with the Net was dull. Front-ends were limited to text-based menus and prompts without the amenities of fonts, graphics, and multimedia. The Web changed all that. Web pages contain graphics, photos, and professional-looking typefaces.

The Web supports hypertext. Hypertext allows a user to click on words or buttons that can take a user immediately to related material. That material may be another document, a photo or graphic, a sound bite, or a video clip. It connects the Web with other Internet resources, such as FTP.

A number of first-class search engines are available for the Web. These engines index what's on the Web.

"Yahoo–A World Wide Web Directory"

http://www.yahoo.com/

While the WWW has generated an enormous amount of excitement, some cautions are in order:

- Fewer than 10 percent of all Internet mail users have the technology and inclination to visit the Web

- Until they're cable modems, the WWW is too slow and too unattractive to attract a truly mass audience

- Fewer than 1 percent of all college students have ever used the Web to shop. In fact, most online shopping experiments haven't been successful. When the Rolling Stones built a widely popular Web page (more than 100,000 hits a day) to sell stuff, the average day saw 3(!) orders.

The widely popular Rolling Stones Web Site

http://www.stones.com/

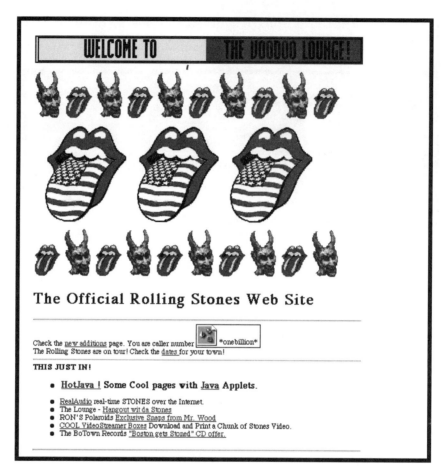

The Internet

 EXPLORING THE NET

INTERNET SEARCH

If you're trying to find a particular site or document on the Internet or just looking for a resource list on a particular subject, you can use one of the many available on-line search engines. These engines allow you to search for information in many different ways - some search titles or headers of documents, others search the documents themselves, and still others search other indexes or directories.

SEARCH ENGINES

INFOSEEK SEARCH
  InfoSeek is a comprehensive and accurate WWW search engine. You can type your search in plain English or just enter key words and phrases. You can also use special query operators:

  [Run Query] [Clear Query Text]

THE LYCOS HOME PAGE: HUNTING WWW INFORMATION
  This search engine, served by Carnegie Mellon University, will allow you to search on document titles and content. Its May 1 database contains 3.75 million link descriptors and the keywords from 767,000 documents. The Lycos index is built by a Web crawler that can bring in 5000 documents per day. The index searches document title, headings, links, and keywords it locates in these documents.

WEBCRAWLER SEARCHING
  This engine allows searches by document title and content. It is part of the WebCrawler project, managed by Brian Pinkerton at the University of Washington, which collects documents from the Web.

SEARCH ENGINE SEARCH

If you still haven't found what you're looking for and you'd like to try out other available search engines, check out these other lists of search engines:

W3 SEARCH ENGINES
  Published by the University of Geneva, this list of search engines covers a wide variety of topics and subjects but isn't updated very often.

CUSI (CONFIGURABLE UNIFIED SEARCH INTERFACE)
  Nexor U.K. offers this tool, a single form to search a large number of different WWW engines for documents, people, software, dictionaries, and more.

Find out more about Netscape at info@netscape.com, or call 415/528-2555.
Copyright © 1995 Netscape Communications Corporation

Various Internet Search Engines

http://home.netscape.com/home/internet-search.html

Present and Future Impact of the Internet

According to the popular media, the Internet is rapidly becoming ubiquitous. While this isn't guaranteed, it's clear that the Net is a media darling, and has a reputation far out of proportion to its size. But because the opinion leaders have adopted it, it's likely to catch on. Today radio stations routinely recite their e-mail addresses for responses from listeners. Cable television networks broadcast Web addresses for their home pages.

"Projections of Internet-related business range to $50 billion at the end of the decade," notes Vinton G. Cerf, senior vice president for the data services division of MCI Telecommunications and co-inventor of TCP/IP. "While this is still small compared to the total telecommunications business—estimated at $300 billion today—its rapid growth and the rich evolution of new products and services suggest that the modest research investments of the federal government have paid off in myriad ways, not all of them merely monetary. There is every reason to believe that the Internet will transform education, business, government, and personal activities in ways we cannot fully fathom. Virtually none of this would have happened as rapidly, or in the same open and inclusive fashion, had not the federal government consciously provided sustained research funding and encouragement of open involvement and open standards, and then wisely stepped out of the picture as the resulting systems became self-sustaining. The Internet is truly a global infrastructure for the twenty-first century—the first really new infrastructure to develop in nearly a century."

3 ONE TO ONE

To really understand the power of the new marketing paradigm, we need to take a look at the current state of marketing.

The two key elements of marketing and business in this century have been mass production and marketing.

Mass Production and Mass Marketing

Mass production was the by-product of the Industrial Revolution. Before mass production, everything was made by hand. The income of a craftsman was directly related to the quantity of goods he could create, and the cost of an item was directly related to how long it took.

There are obvious difficulties with custom production, the two most obvious being that it's extremely difficult to make a lot of money, and goods tend to be expensive.

Henry Ford, assembly lines, the steam engine, and the other innovations of the late 1800s changed that. Suddenly, the investment to be made was in a production system, not in individual items. The cost of the factory, setup, equipment, and training could be huge. But once the factory was in place, the cost of each item produced was the same.

The profit per item increased as production increases. In other words, the more you made, the lower the total cost per item and the higher the profit margin.

Ford realized this and embraced it. He decided to produce as many Model A cars as possible, driving his competition out of business by lowering the price as he went. If a Ford plant was producing twice as many cars as a neighboring car plant, Ford was able to amortize his fixed costs over more cars, allowing him to sell his cars for a retail price that was lower than his competitors' costs.

The engine for mass production was in place. Economies of scale demanded that companies produce as many items as possible in order to drive down costs and thus drive down prices.

The desire for mass sales of mass-produced items led to a demand for mass marketing—in a nutshell, reaching the largest number of people with the lowest possible cost and encouraging them to buy something.

Mass marketing is indiscriminate. A women's shoe maker is willing to advertise in *Time* magazine, even though he knows that half the magazine's readers will never buy a pair of his shoes. Why? Because the CPM—the cost per thousand—is low enough to make the ad an efficient one.

CPM is the mantra of the mass marketer. It puts a dollars-and-cents value on reaching 1,000 potential customers. During the glory days of mass marketing, a brand manager merely compared the CPM for different magazines, TV shows, and other outlets and assembled the most efficient advertising schedule.

Of course, it was never quite that simple. The demographics of the reading audience had to be considered. *Sports Illustrated* was never a good place to advertise diapers, and daytime TV is a dud for hair replacement for men. But demographics can only go so far. In a classic paper, "Are Grace Slick and Tricia Nixon Cox the Same Person?", John O'Toole, an ad exec, pointed out that the counterculture rock star and the President's daughter were the same age, had approximately the same income, and were of the same gender. Yet they obviously were very, very different people.

In a growing, largely homogeneous culture, mass marketing is almost a business panacea. The great marketing companies of our society all grew mighty during this era. Procter & Gamble, Chevrolet, Fruit of the Loom,

Fisher Price, Kodak, Dole, American Tourister Luggage, Spiegel, Marlboro, and other "household names" built their profits and their reputations by mass marketing.

Mass marketing and mass production proved to be a formidable competitive weapon. Who could unseat Jell-O, with its hundreds of millions of dollars in sales, shelf space in every supermarket, efficient plants, and loyal customers? What chance would any competitor have to unseat the leader.

No era lasts forever, though, and we're seeing the unraveling of the age of mass. There were several Achilles' heels:

- Focus on the factory, not on the consumer

- Lack of choice

- Increasing decline in mass-marketing efficiency

- Birth of the generic

Let's look at each. The age of mass, almost by definition, is about the factory. Get the yield up, sell enough to keep the shifts moving, and profits will follow. But by focusing on the factory, not the consumer, the marketer leaves herself open to several threats. The first is an almost inevitable decline in quality.

In a closed market with few choices, a producer can always cut a few corners to increase profits. The consumer, trained to continue purchasing the same brand without questioning quality, will oblige the producer by ignoring the first few defects. Sooner or later, however, a quality backlash erupts and the mass marketer is hurt.

The most obvious example is Detroit. There were three major car companies, and in the cloistered world of southern Michigan, they eventually realized that if they all made the cars a little faster and a little cheaper, huge profits would follow.

Japan manufacturers capitalized on this opening. They offered better cars for less money. It took more than a decade to undo the huge lead that Detroit had built through mass marketing, but with persistence, the Japanese won.

This segues into the second element, lack of choice. If you're trying to make a factory more efficient, one way to do it is to limit the options. "Any color, so long as it's black" is more than a cliché. It's the mantra of any true mass marketer. Choices lead to stocking problems, production problems, advertising problems. Gold Toe socks offered more than 300

Presenting Digital Cash

> **Take a look at almost any category of consumer product. Beer? The popular brand isn't Bud or Miller. It's "other." Same with computers, sneakers, jeans, beverages and TV networks. The days when someone could capture an absolute majority of a market are gone (with the possible exception of Microsoft!).**

different colors and models, and that turned into a management nightmare worthy of a Harvard Business School case. Choice was the enemy. The goal of the mass marketer was to narrow down the bulk of the market to the smallest possible number of choices.

Limited choice works as long as supply is limited. For example, when we bought our office supplies at a tiny local store, it's unlikely we noticed that Post-It notes only came in one color. But once Staples and other office superstores opened their doors, there was plenty of room for choice. So competitors walked in with new sizes, new colors, new choices. And the hegemony started to break down.

People want choice. As Americans became more affluent, they also became choosier. The average supermarket now stocks more than 30,000 items. And there are another 17,000 new products introduced every year. Ten years ago you could choose from three kinds of premade iced teas at your local supermarket. Now there are more than 250 on the market.

Take a look at almost any category of consumer product. Beer? The popular brand isn't Bud or Miller. It's "other." Same with computers, sneakers, jeans, beverages, and TV networks. The days when someone could capture an absolute majority of a market are gone (with the possible exception of Microsoft!).

The third factor in the eventual decline of mass marketing was just too much of a good thing. For decades, magazines, newspapers, radio, and TV drank deeply from the river of money spent by mass marketers on advertising. It's not unusual for a national advertiser to spend $100 million in a year, trying to reach consumers.

But as more companies jumped on the mass-marketing bandwagon, consumers' minds began to feel the glut. The average American sees more than 100,000 advertising messages a year. By the time a child is six years old, she's already seen more than 50,000 commercials for toys or cereal.

Fill in the blank: Winston tastes good...

If you're over thirty, chances are you know the answer. Yet this jingle hasn't been broadcast since 1969! Mass marketing, in a relatively uncluttered market, was able to burn it into your mind. Today, performing the same task would be far, far more difficult.

Advertising is ceasing to be a panacea. A nationwide TV rollout no longer guarantees a stream of customers walking into the store and asking for a product. Gaining the attention of increasingly jaded consumers is proving to be increasingly expensive and increasingly difficult.

The obvious result of these three factors is that mass marketing just isn't working as well as it used to. As the market fragments, producers can't offer their goods as cheaply. And the brand image created by marketers hasn't proved resilient enough to guarantee the effectiveness of price increases.

When there were only five cereals to choose from, Kellogg's was able to price Corn Flakes low enough to eliminate competition in their category. The huge sales helped offset the mass advertising costs, and the people in Battle Creek were quite happy. Today, with a simple box of Corn Flakes retailing for more than $3, consumers have responded by embracing private labels and generics.

This is truly alarming to the makers of mass-marketed products. It means that a few hundred million dollars in advertising, spent judiciously over a fifty-year period, accompanied by great packaging and the best sales force in the business, is not enough to overcome a 75-cent price advantage to the generics. It means the time, the money, and the effort being expended in mass marketing are no longer as useful as they were.

This realization is shaking mass marketers to their roots. It is a cause for major concern among all purveyors of media, as well as distributors and factory workers. If price becomes the primary differentiator among goods, the only winner will be the consumer.

Stop for a minute and imagine what it would be like to have been born in Swaziland, or New Zealand, or Szechuan province. Can you imagine the upbringing, education, cultural experiences, and life you would have had? Probably not. It's so far removed from your experience, so different from what you've come to expect, that the leap is too difficult to make. The paradigm shift discussed in this book is similar, though perhaps not as dramatic.

What if we posit that the future will not be about mass (mass production and mass marketing) but about individuals? What if the goal of every marketer is not to maximize market share and factory output, but to grab and keep the largest share of each individual consumer? What if car dealers, florists, sneaker makers, and restaurants all competed to discover exactly what you want, and strove to please you?

Welcome to the one-to-one future. Don Peppers and Martha Rogers wrote one of the best business books of the decade, *The One to One Future*. In this classic, they outlined how marketing will change over the next decade. A few key elements that will characterize this future:

First, the focus will be on customer share, not market share. In our overmarketed society, the largest cost is introducing yourself to a customer, building the trust necessary to do business. Once that

As the market fragments, producers can't offer their goods as cheaply. And the brand image created by marketers hasn't proved resilient enough to guarantee the effectiveness of price increases.

> "Share of customer" only works if there is a one-to-one communication channel available between producers and consumers. And until recently there hasn't been one. The brand manager of Jell-O has absolutely no idea if you like Jell-O. She doesn't know what flavors you like, nor does she know what else you like for dessert.

introduction is built, both the marketer and consumer benefit if they can maximize the business they'll do together, rather than taking the time to shop around for substitutes.

Imagine that a bank spends $200 in marketing to attract a new customer. Once that customer is in the door and satisfied, though, the cost of bringing new services to that customer is quite low. Investments, insurance, even personal accounting services are all sensible things to offer. From the consumer's point of view, what's the incentive to switch? If the bank, which knows the consumer's finances anyway, can provide additional services that leverage that consumer's time and energy, more power to it!

Unfortunately, most banks were oblivious to this calculation. Even worse, when they try to implement these additional services, they're often treated as ill-fitting add-ons, not integrated into the service package being offered.

This works outside the service arena as well. Levi Strauss, for example, has long wanted more than just the weekend clothing business of its customers. After an ill-fated attempt to produce suits, they created their hot-selling Dockers line of Fridaywear. They then went out and persuaded businesses to improve their productivity by *permitting* Friday wear. They created a new market, and worked hard to satisfy their existing customer base. If you were happy with your Levi's, why switch?

"Share of customer" only works if there is a one-to-one communication channel available between producers and consumers. And until recently there hasn't been one. The brand manager of Jell-O has absolutely no idea if you like Jell-O. She doesn't know what flavors you like, nor does she know what else you like for dessert. It's virtually impossible for General Foods (now Kraft) to increase its share of consumer because they have no idea whom they sell to. They sell to the "mass," not the individual.

E-mail, the Internet, and toll-free phone numbers dramatically increase the opportunity for businesses to establish relationships with consumers. Add databases and computers into the mix, and businesses can now keep track of every single customer for an annual cost approaching that of two TV commercials.

How does this communication help a marketer increase share of consumer? Let's imagine you told Macy's that you were pregnant. Macy's now knows a great deal about your needs over the next three years, and can work to satisfy them.

At four months, it can send you a letter inviting you to a private showing of the latest maternity clothes (priced at a sizable discount, since they didn't have to pay for a full-page ad to let you know about their maternity department). Four weeks before the baby's birth you might get a call from someone in the layette department, asking if you'd like them to set aside all the clothes and linens you'll need after delivery, thus saving you the time and hassle of trekking to the store.

Two months later, when it's time for a blowout baby shower, Macy's might be happy to let you register for baby gifts via e-mail. And they can even offer you a subscription service, sending clothes to your house on approval—always in just the right size, of course. And since Macy's "owns" a relationship with you, it can leverage this by doing a deal with Random House that offers you a book every three weeks (age appropriate, of course).

Invasion of privacy? Not really. After all, you volunteered the information. You're saving time and money and avoiding the hassle of finding new stores, returning things that don't fit, and getting three sets of the same outfit. And you can stop the process at any time.

So far, the process described is one-way. You initiate it; then Macy's does all the talking. But in the one-to-one future, the communication is definitely two-way. Encouraging customers to make suggestions, complain, and speak out is essential.

Research shows that a satisfied customer generally talks to one other person about the experience. An unhappy customer tells nine. Obviously the cost of unhappy customers is huge. By making it easy to complain, marketers are able to short-circuit the unhappy-customer cycle every once in a while, even turning unhappy customers into missionaries, eager to talk about their great experience.

But dealing with complaints is not just a defensive maneuver. Complaints and customer feedback are the loop that can produce far better products. Marketers need this feedback to shorten the development loop for new products and services and to better gauge the needs and desires of their customer base.

American Express is a great example of a company that hasn't completed its feedback loop. Once dominant in charge cards and traveler's checks, the company is in danger of being marginalized because it didn't listen to consumers. If they had heard the desire for revolving credit fifteen years ago, they could have cut Visa and MasterCard off at the pass.

Obviously, e-mail and databases are the single best way to complete this loop. For years, Bill Gates answered every piece of e-mail he

> **Let's imagine you told Macy's that you were pregnant. Macy's now knows a great deal about your needs over the next three years, and can work to satisfy them.**

received. Corporations can build auditable, archivable records of consumers' wants and beefs, and even better, can respond to them directly. The packaged-goods companies and other mass marketers just aren't equipped to deal with this.

Try calling the president of Kellogg's to complain about a product. Good luck. Want to share a new idea with Phil Knight at Nike? You'll probably get a letter back from their lawyers. Mass marketers have built a moat around their fortress, keeping consumers as far away as possible. Contrast this with Steve Case, CEO of America Online. Steve spends hours reading e-mail from his subscribers, and it's not surprising to see a consumer suggestion turn into a service feature, sometimes within weeks.

Mass marketers have absolutely no idea what to do with individuals. That's why they're mass marketers. The new marketing paradigm calls for a one-to-one relationship, and that requires computers to track and cater to each individual.

One of the first steps is to differentiate customers—to group customers not just by standard demographic data, but by common interests, wants, and desires. Grace Slick is not Tricia Nixon, but she does have a lot in common (politically) with Jerry Brown, and fashion-wise with every other aging baby boomer. **Understanding the true meaning behind the obvious demographics is an excellent place for a reformed mass marketer to start.**

But to gather this information, marketers need a simple, inexpensive two-way communication mechanism. Consumers will gladly share information if they believe that the sharing will lead to benefits. To date, most consumer data collection has been clunky at best, useless at worst. Those little warranty cards you fill out when you buy a stereo or a dishwasher are a great example—many companies box them up and never even use the data. The ones that do use the data find that it's extraordinarily expensive and time consuming to enter. Often, by the time a marketer has access to the data on his computer, the consumer has moved, outgrown the product, or died.

The Internet offers an instant, virtually free solution to this problem. Using e-mail, it's a matter of seconds for a consumer to enter some data, and a matter of days, not months or years, before the data appear on the marketer's computer database.

Once the connection has been made, marketers can start worrying about economies of scope, as opposed to economies of scale. What's that mean? An economy of scale is an advantage gained by ramping up production. It's much cheaper to make a million sneakers at once than it is to build them a few at a time. Traditional marketing, as previously discussed, has focused on this economy.

But the real payoff is in economies of scope. Basically, once you have a customer, you should take advantage of that relationship and produce as many products and services for that customer as your niche and expertise permit you to. Once you accept that acquiring a customer and building trust are the time-consuming, expensive tasks, it becomes clear that the most highly leveraged activity for a marketer is to sell more products to existing customers.

A garden store, for example, should note that all of its customers need shovels, hoses, underground sprinklers, outdoor furniture, barbecues, fences, and more. There's no reason that this store can't establish itself as an outdoor clearinghouse, offering its harried customers an opportunity to buy an entire lifestyle from one source.

In the electronic world, this becomes even more clear. Subscribers to one newsletter are ideal customers for subscriptions to a newsletter on a related topic. And satisfied customers are far more likely to buy an incremental newsletter from the same publisher.

Now we're ready to turn the equation totally upside down. If marketers are focusing on economies of scope, trying desperately to bring new consumers into the fold and keep them, if we accept that two-way, one-on-one communication is the basis for long-term profitability, we're led to a buying system that starts at the consumer and moves to the producer.

Namely, the consumer identifies his need, outlines it to a number of producers, and waits for bids. For example, if you're eager to buy a new car, you could post a note to every dealer in town. Outline the budget you have in mind, the features you need, even the models you're interested in. Then sit back and let the network take care of you.

Your automated robot can automatically contact all the dealers who have shown an interest in acquiring new customers. These dealers will receive your notice, then hustle to find exactly what you need. Even better, the message could go straight to the manufacturer, who could offer to build a car to your specifications!

Far-fetched? Not really. The manufacturer/retailer has already established an infrastructure that supports this sort of searching, but it requires you to get in your car and spend days driving around, having inefficient conversations over and over again. Didn't like the sport coat that salesman was wearing? You might have walked out because of that, even though it will have no effect on the car you'll eventually buy.

One of the dealers you contact might evaluate your needs and suggest a car you hadn't even considered. This value-added feature in the shopping process is far easier to supply by e-mail, and it offers benefits to both you and the dealer.

The Role of Digital Cash in This Paradigm

Mass marketers sowed the seeds of their system's destruction when they continued to up the ante in their advertising. The cost of gaining share of mind continues to increase, and an increasingly jaded consumer is getting more and more difficult to reach.

The market has been waiting for a more efficient, more rational model. That model is starting to fall into place. Databases to manage information are here. Addressable marketing techniques, from direct mail to e-mail, now allow individual communication. The missing link is a way for consumers to make payments (both large and small) and an easy way for marketers to collect money from individuals.

Traditionally, retailers have been best equipped to deal with revenue collection from large numbers of people. Manufacturers and service providers have often shied away from this portion of the chain, even though the margins here are significant. Yet if a marketer is going to invest in the expense of establishing one-to-one relationships with consumers, she almost always has to remove the retailer (and the retailer's high margins) from the chain in order to profit.

Digital cash makes this last step possible. Nestlé can offer a Quik flavor-of-the-month club to its best customers, without handing the bulk of its margin over to a fulfillment house that processes handwritten forms, cashes checks, and deals with charge cards. Even better, consumers can find themselves making small purchases without the shipping and handling charge dwarfing the purchase itself.

Garry Trudeau can create a link with every one of *Doonesbury's* 5 million loyal readers. But instead of going through the tedious process of dealing with a syndicator who deals with newspapers who deal with newsstands who deal with consumers, Trudeau can sell his cartoons directly to the public. At 1 cent a person, Trudeau will see his profits skyrocket. Consumers will benefit from direct interaction with Trudeau and his database. If a lot of people loved his Frank Sinatra series, Trudeau can choose to visit the topic again. Under the current system, however, he has no idea what people like and what they don't.

CRYPTOLOGY

Before we get into the nuts and bolts of digital cash, it's important to understand something about cryptology. "Cryptography—technology which allows encoding and decoding of messages—is an absolutely essential part of the solution to information security and privacy needs in the Information Age," Electronic Frontier Foundation executive director Jerry J. Bermanis told a congressional committee in 1994. "Without strong cryptography, no one will have the confidence to use networks to conduct business, to engage in commercial transactions electronically, or to transmit sensitive personal information."

Think of digital cash as the serial number on a dollar bill that has shed its paper body and entered the ether.

You see, digital cash is essentially a number. Think of it as the serial number on a dollar bill that has shed its paper body and entered the ether. The paper allows the number to be passed around the economy. It also gives the number a measure of protection from counterfeiters—although that measure has been shrinking as low-cost printing

45

technologies have become increasingly sophisticated. You can photocopy a dollar bill, but you won't get much farther than the first round of drinks in a bar when you try to pass it off as money. You can make plates and try to print your own money, but that's an expensive proposition. We know millions of dollars in "queer" money is in circulation, but making counterfeit money isn't easy. Knocking off digital money should be as hard, if not harder. That's where cryptology comes in. It makes digital cash, or the paperless serial number, as secure as its paper counterpart.

What Is Cryptology?

There's a tale about Julius Caesar. They say he didn't trust his messengers, so when he sent important messages to his associates, he first encoded them. He would replace an A with a D, a B with an E, and so on. To decipher the message, the recipient would have to know the "key," which was "shift by three."

Caesar's simple scheme is what's called a crypto- or ciphersystem. It's a way to disguise messages to limit who may read them. The art of creating and using these systems is called cryptography; the art of breaking them is called cryptanalysis. Cryptology is the study of both.

Caesar's original message—the one that could be read by any friend, Roman or countryman—is called a plaintext. Caesar converted his plaintext into ciphertext, a disguised message, through encryption, which is any procedure that converts plaintext into ciphertext. When the message arrived at its destination, its recipient would decrypt it, or convert the ciphertext back into plaintext.

A cryptosystem is usually a collection of algorithms. In Caesar's case, the algorithm was "shift by n," where n equals 3. If n were always equal to 3, Caesar's system would be pretty easy to crack, so to make things a little harder, Rome's first emperor probably varied the value of n. Sometimes it might be 3; other times 4; and so forth.

In a digital cash system, then, the number of each unit needs to be encrypted, or disguised, so it won't be misappropriated.

Public Key Cryptography

Caesar's system is an example of secret key cryptography. Only Caesar and the recipient of the message have the key to decrypt the message. For secret key cryptography to be effective, the cryptographic key must be kept secret and controlled only by those who have access to the key. In a digital cash system, public key cryptography is used.

Public key cryptography makes use of two keys: a public key and a private key. The two keys are mathematically related, but the private key cannot be determined from the public key. In a system implementing public key technology, each party has its own public/private key pair. The public key can be known by anyone; however, no one should be able to modify it. The private key is kept secret. Its use should be controlled by its owner, and it should be protected against modification as well as disclosure.

Being able to send secure messages is one function of public key cryptography. Another function is authentication, or verifying that the message came from who it says it came from. For example, if Joe sends you ten cyberbucks, you want to make sure they came from Joe and not Louie, who doesn't have ten cyberbucks to his name.

Quadralay's Cryptography Archive–links to all sorts of other Web pages and documents about cryptography

http://www.quadralay.com/www/Crypt/Crypt.html

Quadralay's Cryptography Archive

Please sign our Guest Book.
You may also Search This Server.

Clipper
Crypto-Anarchists
    Overview
Cypherpunks
    Berkeley Archive
    CMU Archive
    University of Washington Archive
Digital Encryption Standard (DES)
    Books with DES Source Code
Digtial Telephony
    EFF Digtal Telephony Archive
Dolphin Encrypt
    Overview
Export Issues
    Cygnus Archive
    EFF Crypto Export Archive
General
International Cryptography Pages
Kerberos
    Frequently Asked Questions
    Reference Page
Macintosh Cryptography Interface Project
National Security Agency (NSA)
    Clautrophobia Article
    Houston Chronicle Interview
    NSA can break PGP (joke)
    Original NSA Charter
    Newsgroup: alt.politics.org.nsa
    Security Guidelines
PGP (Pretty Good Privacy)
    EFF PGP Archive (info, no code)
    Frequently Asked Questions
    MIT Distribution Site
    Newsgroup: alt.security.pgp
RIPEM
    Archive at Indiana University
    Frequently Asked Questions
    Frequently Noted Vulnerabilities
    Newsgroup: alt.security.ripem
RSA
    Frequently Asked Questions
    RSADSI Web Server
Sternlight, David
Tempest
    Overview

Information contained in this cryptography archive does not neccessarily represent the views of Quadralay Corporation. It is provided for educational purposes only.

Digital Signatures

One way to authenticate a message is through a digital signature. A digital signature is a computer-based method of "sealing" an electronic message in such a way that its contents cannot be changed or forged without detection and that the identity of the originator of the communication can be verified.

The digital signature for a message is simply a code, or large number, that is unique for each message and each message originator (within a very high, known probability). A digital signature is computed for a message by computing a representation of the message (called a hash code) and a cryptographic process that uses a key associated with the message originator. Any party with access to the public key, message, and signature can verify the signature. If the signature is verified, the receiver (or any other party) can have confidence that the message was signed by the owner of the public key and the message has not been altered after it was signed.

To generate a signature on a message, the owner of the private key first applies a secure hash algorithm to the message. This action results in a condensed representation of the message, known as a message digest. The owner of the private key then applies the private key to the message digest. Sound complicated? Not really. Basically, a key modifies the body of the message, making it unreadable. Any party with access to the public key, message, and signature can verify the signature using a standard digital signature algorithm, an authentication algorithm for digital signatures. The most common digital signature algorithm used today is the one developed by RSA.

Public keys are assumed to be known to the public in general. If the signature is verified, the receiver (or any other party) has confidence that the message was signed by the owner of the public key and the message has not been altered after it was signed.

In addition, the verifier can provide the message, digital signature, and signer's public key as evidence to a third party that the message was, in fact, signed by the claimed signer. Given the evidence, the third party can also verify the signature. This capability, an inherent benefit of public key cryptography, is called nonrepudiation.

The digital signature does not provide confidentiality for information in the message. It only establishes that the recipient signed the message. If confidentiality is required, the signer needs to encrypt the message.

The Digital Encryption Standard

One standard for encrypting messages is the Digital Encryption Standard, or DES.

The standard's origins lie in an encryption standard for unclassified government computer data and communications, called Lucifer, developed by IBM in the early '70s. It was certified by a civilian agency, the National Bureau of Standards (now NIST), as the Data Encryption Standard in 1976.

Unlike public key cryptography which uses two keys (either one of which may be used to encrypt, and the other to decrypt), DES is a secret key, symmetric cryptosystem: When it is used for communication, both sender and receiver must know the same secret key, which is used both to encrypt and decrypt the message. DES can also be used for single-user encryption—for example, to store files on a hard disk in encrypted form. In a multi-user environment, secure key distribution may be difficult; public key cryptography was invented to solve this problem.

DES operates on 64-bit blocks with a 56-bit key. It was designed to be implemented in hardware, and its operation is relatively fast. It works well for bulk encryption—that is, for encrypting a large set of data.

DES has never been "broken," despite the efforts of many researchers over many years. The obvious method of attack is brute-force exhaustive search of the key space.

What is a brute-force search and what is its cryptographic relevance? In a nutshell: If $f(x) = y$ and you know y and can compute f, you can find x by trying every possible x. That's a brute-force search. For example: Say a cryptanalyst has found a plaintext and a corresponding ciphertext, but doesn't know the key. He can simply try encrypting the plaintext using each possible key, until the ciphertext matches—or decrypting the ciphertext to match the plaintext, whichever is faster. Every well-designed cryptosystem has such a large key space that this brute-force search is impractical. But advances in technology sometimes change what is considered practical. For example, DES, which has been in use for over ten years now, has 2^{56} possible keys. A computation with this many operations was certainly unlikely for most users in the mid-'70s. The situation is very different today, given the dramatic decrease in cost per processor operation. Massively parallel machines could threaten the security of DES against a brute-force search. It has been suggested that a rich and powerful enemy could build a special-purpose computer capable of breaking DES by exhaustive search in a reasonable amount of time.

> **In a multi-user environment, secure key distribution may be difficult; public key cryptography was invented to solve this problem.**

But the attack would be costly, around $1 million.

However, the consensus is that DES, when used properly, is secure against all but the most powerful enemies. In fact, triple-encryption DES may be secure against anyone at all. It is used extensively in a wide variety of cryptographic systems, including electronic funds transfers between banks; in fact, most implementations of public key cryptography include DES at some level.

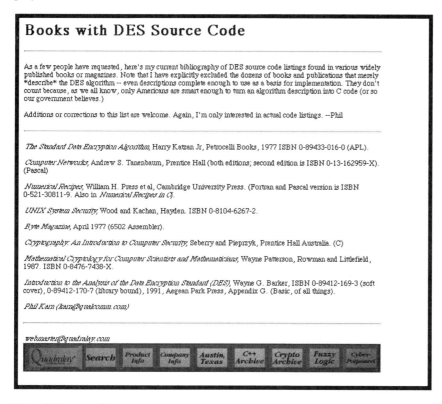

A listing of books with DES source code

http://www.quadralay.com/www/Crypt/DES/source-books.html

The Clipper Controversy

Cryptography has become a bone of contention between the United States government and the nation's high-technology industry. This wrestling match has significant consequences for electronic commerce in general and digital cash in particular. For example, one reason cited by DigiCash for locating its headquarters overseas is restrictive U.S. policies on the export of strong cryptographic schemes.

In the United States, cryptography has long been the personal pond of intelligence spooks, most notably the National Security Agency

(NSA). As weighty as the concerns of the NSA may be over the proliferation of cryptographic schemes that can be used by the "bad guys" to mask criminal activity, the agency's motives are suspect in the eyes of many people interested in encryption issues. These people reason that, like the Central Intelligence Agency, which is looking for something to occupy the idle hands of its spymasters now that the final nails have been hammered into the coffin of the Evil Empire, the NSA needs a cause to justify its existence. The cause it has chosen is the "Clipper chip."

The government sees Clipper as a vehicle for balancing its interest in combating crime and the citizenry's interest in privacy. It also illustrates how fearful established institutions can become in the face of digital anarchy. For years, law enforcement agencies have tapped telephones to gather evidence against all sorts of lowlifes like Mafiosi, terrorists, and so forth, and against not-so-low lowlifes like political activists and enemies of Richard Nixon. Now that the telephone system is becoming more and more digital, more and more conversations will be singing through fiber optic cable in the form of zeroes and ones. In that form, the conversations can be encrypted. And if they're encrypted, law enforcement agencies won't be able to intrude on them.

So the government cooked up the idea of Clipper, a semiconductor device installed in telephones and modems that encrypts digital voice and data transmissions and interprets communications that have been encrypted with the same algorithm. We've reviewed public key systems and secret key systems. Clipper is what's called an escrow key system because the keys for decoding a user's conversations are held by a third party "in escrow." In this case, the decoding key necessary to interpret communications that have been encrypted with Clipper would be held by both the Treasury Department and NIST. Any law enforcement agency granted a court order to intercept communications that have been encrypted with the chip would first have to obtain the decoding key from Treasury and NIST before it could execute the authorized surveillance. Having two agencies safeguarding the keys adds a security level to the scheme. The thinking is that it is twice as hard to filch the keys from two agencies than from one.

In theory, this is how the law enforcement agencies would use Clipper:

The FBI suspects you're involved in a conspiracy to distribute child pornography. It wants to tap your digital phone, which is Clipper chip enabled. The G-men go to a judge, persuade him they have a good reason to tap your phone, and obtain a warrant. They present the warrant to the keepers of the keys to your phone. The keepers turn the keys over

> **Every well-designed crypto-system has such a large key space that this brute-force search is impractical. But advances in technology sometimes change what is considered practical.**

to the agents, and they tap your phone and decode your conversations.

When Clipper was announced in February 1993, it set off a bigger buzz than a bear paw in a beehive. High-tech industry leaders condemned the approach because they believed it gave a competitive edge to overseas outfits who would not be cinched into an encryption-standard straitjacket by their governments. Encryption experts knocked the approach because they argued it could easily be circumvented. Civil libertarians rapped the thin privacy protections in the scheme. From other, maybe more paranoid quarters came questions about where Clipper originated.

The encoding scheme for Clipper, known as Skipjack, came from the NSA, which refused to reveal the algorithm behind it. Among cryptographic professionals, it is considered poor form to hide the algorithm used to perform the encryption. This is because encryption algorithms that look good to one cryptographer can be found to be obviously weak by another. By encouraging public review of an encryption method, cryptographers benefit from the experience of other cryptographers, who can more adequately examine an algorithm for weaknesses.

But to conspiracy theorists the NSA's insistence on keeping Skipjack's algorithm in the shadows indicated something more than just a violation of good form. It suggested the NSA had built a "backdoor" into Clipper's design. Whoever had the key to the backdoor could decode whatever was encrypted by any Clipper chip and so had carte blanche to tap any phone anywhere anytime. Who had the key to the backdoor? The NSA, which wants to make sure that in all things cryptographic, it remains king. In various forums—congressional hearings, magazine articles, online newsletters—the NSA denied it had built in a backdoor to Clipper.

Stewart A. Baker, chief counsel for the NSA, defended Clipper in an article appearing in *Wired* magazine. In the article, Baker attempted to dismiss what he called seven "myths" about Clipper, which he said was 16 million times stronger than DES—the existing standard used, among others, by the government and financial institutions.

Baker argued that Clipper would not be a "willy-nilly" intrusion into the privacy of citizens. "All that key escrow does is preserve the government's current ability to conduct wiretaps under existing authorities," he wrote. "Even if key escrow were the only form of encryption available, the world would look only a little different from the one we live in now. In fact, it's the proponents of widespread unbreakable encryption who want to create a brave new world, one in which all of us—crooks included—have a guarantee that the government can't tap our phones. Yet these proponents have done nothing to show us that the new world they seek will really be a better one."

The NSA lawyer contends that wiretapping is preferable to other means of corralling crooks such as flipping co-conspirators. "I once did a human rights report on the criminal justice system in El Salvador," Baker recalled. "I didn't expect the Salvadorans to teach me much about human rights. But I learned that, unlike the U.S., El Salvador greatly restricts the testimony of 'turned' co-conspirators. Why? Because the co-conspirator is usually 'turned' either by a threat of mistreatment or by an offer to reduce his punishment. Either way, the process raises moral questions—and creates an incentive for false accusations."

On the other hand, wiretaps help convict criminals with their own uncoerced words. "In addition," Baker continued, "the world will be a safer place if criminals cannot take advantage of a ubiquitous, standardized encryption infrastructure that is immune from any conceivable law enforcement wiretap. Even if you're worried about illegal government taps, key escrow reinforces the existing requirement that every wiretap and every decryption must be lawfully authorized. The key escrow system means that proof of authority to tap must be certified and audited, so that illegal wiretapping by a rogue prosecutor or police officer is, as a practical matter, impossible."

Another myth advanced by Clipper opponents, according to Baker, is that unreadable encryption is the key to our future liberty. "This sort of reasoning is the long-delayed revenge of people who couldn't go to Woodstock because they had too much trig homework," he declared. "It reflects a wide—and kind of endearing—streak of romantic high-tech anarchism that crops up throughout the computer world.

"The problem with all this romanticism is that its most likely beneficiaries are predators," he contends. "Take for example the campaign to distribute PGP ["Pretty Good Privacy"] encryption on the Internet. Some argue that widespread availability of this encryption will help Latvian freedom fighters today and American freedom fighters tomorrow. Well, not quite. Rather, one of the earliest users of PGP was a high-tech pedophile in Santa Clara, California. He used PGP to encrypt files that, police suspect, include a diary of his contacts with susceptible young boys using computer bulletin boards all over the country."

A third misconception by Clipper opponents identified by Baker is that encryption is the be-all and end-all of security in a digital world. In reality, encryption is only a small part of network security, he wrote.

Information that citizens willingly release—information on mortgage applications, credit card forms, and magazine subscriptions—poses a greater threat to privacy than anything that would be gathered through the Clipper system. Keeping that information secure is a much more weighty issue than Clipper, and one to which encryption won't provide a resolution.

But Clipper opponents also argue that the government's claim that Clipper is "voluntary" is a ruse. The real agenda is to make key escrow mandatory. Baker contends that those notions are wrongheaded. He wrote: "What worries law enforcement agencies—what should worry them—is a world where encryption is standardized and ubiquitous: a world where anyone who buys an US$80 phone gets an 'encrypt' button that interoperates with everyone else's; a world where every fax machine and every modem automatically encodes its transmissions without asking whether that is necessary. In such a world, every criminal will gain a guaranteed refuge from the police without lifting a finger.

"The purpose of the key escrow initiative is to provide an alternative form of encryption that can meet legitimate security concerns without building a web of standardized encryption that shuts law enforcement agencies out," he continued. "If banks and corporations and government agencies buy key escrow encryption, criminals won't get a free ride. They'll have to build their own systems—as they do now. And their devices won't interact with the devices that much of the rest of society uses. As one of my friends in the FBI puts it, 'Nobody will build secure phones just to sell to the Gambino family.' "

A fifth myth cited by Baker is that Clipper will impede private enterprise from meeting the encryption needs of a free marketplace. Just the opposite may be true, he wrote. The private sector may prefer key escrow to other encryption schemes. "Why?" he asks. "Because the brave new world that unreadable encryption buffs want to create isn't just a world with communications immunity for crooks. It's a world of uncharted liability. What if a company supplies unreadable encryption to all its employees, and a couple of them use it to steal from customers or to encrypt customer data and hold it hostage? As a lawyer, I can say it's almost certain that the customers will sue the company that supplied the encryption to its employees. And that company in turn will sue the software and hardware firms that built a 'security' system without safeguards against such an obvious abuse. The only encryption system that doesn't conjure up images of a lawyers' feeding frenzy is key escrow."

Up to now, Baker explained, the federal government has dominated the encryption market. Development of encryption systems is expensive, so government financing of encryption R&D has been a necessity. That might not be true in the future, but one can't ask the government to pay for the development of encryption schemes that will be used to further activity the government is pledged to stop. The government has proposed a scheme—Clipper—that it feels balances its interests with its citizens' interests. "So where does this leave industry, especially those companies that don't like either the 1970s-vintage DES or key escrow?" he asks. "It leaves them where they ought to be—standing on their own two feet. Companies that want to develop and sell new forms of unescrowed encryption won't be able to sell products that bear the federal seal of approval. They won't be able to ride piggyback on federal research efforts. And they won't be able to sell a single unreadable encryption product to both private and government customers."

Then there's the old NSA chestnut: that the agency shouldn't be involved in this issue because its presence makes any proposal suspect. "With code breakers and code makers all in the same agency, NSA has more expertise in cryptography than any other entity in the country, public or private," Baker noted. "It should come as no surprise, therefore, that NSA had the know-how to develop an encryption technique that provides users great security without compromising law enforcement access. To say that NSA shouldn't be involved in this issue is to say the government should try to solve this difficult technical and social problem with both hands tied behind its back."

Finally, there's the secrecy myth: that the entire Clipper initiative was conducted in the shadows. "This is an old objection, and one that had some force in April of 1993, when the introduction of a new AT&T telephone encryption device required that the government move more quickly than it otherwise would have," Baker explained. "Key escrow was a new idea at that time, and it was reasonable for the public to want more details and a chance to be heard before policies were set in concrete. But since April 1993, the public and industry have had many opportunities to express their views. The government's computer security and privacy advisory board held several days of public hearings. The National Security Council met repeatedly with industry groups. The Justice Department held briefings for congressional staff on its plans for escrow procedures well in advance of its final decision. And the Commerce Department took public comment on the proposed key escrow standard for 60 days."

All that public exposure Baker refers to had an impact with which he may have not been pleased. Clipper was put on the shelf. In July 1994, Vice President Al Gore, in a letter to Washington Democratic representative Maria Cantwell, made it clear that while Clipper might have a small place in the telephone security market, it has no future in the digital world. "The Clipper Chip is an approved federal standard for telephone communications and not for computer networks and video networks," he wrote. "For that reason, we are working with industry to investigate other technologies for those applications.... We welcome the opportunity to work with industry to design a more versatile, less expensive system. Such a key escrow system would be implementable in software, firmware, hardware, or any combination thereof, would not rely upon a classified algorithm, would be voluntary, and would be exportable."

At the time, Gore's letter brought a sigh of relief from Clipper opponents, but their comfort was short-lived. A little more than a year later, the government rebirthed Clipper. And threw gasoline on the all-but-extinguished embers of the key escrow debate. This time a "commercial" entity instead of a government agency would hold a key to the chip's encryption. At this writing the issue remains unresolved.

The development of digital cash can't be viewed outside the Clipper debate. The government's concerns over the abuse of encryption to further criminal activity can be extended to concerns over cyberbucks contributing to tax evasion, money laundering, and illegal gambling. If it becomes kosher to grant the government the power to eavesdrop on digital communications, it's plausible that government could intercept e-cash transactions.

Netscape

Whether or not an encryption scheme can be cracked is an important issue for the proliferation of digital cash. Although cyberbuck advocates feel confident that existing encryption schemes make digital cash safer than paper money, an incident in 1995 involving Netscape has raised some questions about all security on the Net.

Many customers now routinely purchase products over the Internet using their World Wide Web browsers, made by Netscape. Netscape Commerce Server software accepts orders submitted from a Netscape browser, which are filled out in an online Web form and transmitted across the net to the server.

Using the Secure Sockets Layer (SSL) protocol, the customer's name, address, phone number, and credit card number are encrypted with the RC4 algorithm developed by Ron Rivest of RSA Data Security. Due to export restrictions placed upon cryptographic software by the U.S. government, the Netscape software most commonly used is limited to encrypting with a key only 40 bits in length.

RSA's FAQ (Frequently Asked Questions) About Today's Cryptography

http://www.rsa.com/rsalabs/faq/faq_des.html#des.html

In the summer of 1995, a person from France posted a news article to the hacker community claiming success at decrypting a single encrypted message that had been posted as a challenge on the Internet sometime on or before July 14, 1994.

What this person did is decrypt one message that was encrypted using the RC4 algorithm and a 40-bit key. He used 120 workstations and two parallel supercomputers at three major research centers for eight days to do so.

As many have documented, including Netscape, a single RC4 40-bit encrypted message takes sixty-four MIPS-years of processing power to break, and this roughly corresponds to the amount of computing power that was used to decrypt the message.

Netscape, responding to the incident, emphasized that the computerist broke only a single message. For him to break another message (even from the same client to the same server seconds later) would require another eight days of 120 workstations and two parallel supercomputers. The work that goes into breaking a single message can't be leveraged against other messages. Every message uses a different encryption key.

The standard way to judge the level of security of any encryption scheme is to compare the cost of breaking it to the value of the information that can be gained. In this case the decrypter had to use at least $10,000 worth of computing power (ballpark figure for having access to 120 workstations and two parallel supercomputers for eight days) to break a single message. Assuming the message is protecting something of less value than $10,000, then this information can be adequately protected with only RC4 40-bit security. For information of greater value, a stronger encryption scheme should used, such as RC4 128-bit, which uses a longer key, 128 bits, and would be tougher to crack.

The information obtained as a result of decrypting this session was someone's name, address, and a list of items they were trying to purchase online. This information is hardly worth the time and expense it took to obtain it. Even if a credit card number had been obtained as a result of this attack the value of that card number would, in most cases, be less than $10,000. Furthermore, a popular site may have many people browsing and significantly fewer people buying, making it extremely difficult to isolate a valuable session with any acceptable probability.

Inside the United States, software can support a range of stronger encryption options, including RC4 128-bit, which is 288 times harder to decrypt—meaning that the computer power required to decrypt such a message would be more than 1 trillion times greater than that used to decrypt the RC4 40-bit message. This means that with foreseeable computer technology it would be practically impossible.

"In conclusion," Netscape stated, "we think RC4 40-bit is strong enough to protect consumer-level credit-card transactions—since the cost of decrypting the message is sufficiently high to make it not worth the computer time required to do so—and that our customers should use higher levels of security, particularly RC4 128-bit, whenever possible. This level of security has been available in the U.S. versions of our products since last April. Because of export controls it has not been available outside the U.S."

When the challenge to break the RC4 40-bit code was issued on the Internet, the French researcher declared he was confident he could crack it. "I knew I would get the result in at most 15 days, with an expected average of eight days," he wrote in a posting on the Internet. "The actu-

> "We think RC4 40-bit is strong enough to protect consumer-level credit-card transactions—since the cost of decrypting the message is sufficiently high to make it not worth the computer time required to do so..."
> Netscape

al time was the same as the expected time because the result was almost exactly in the middle of the search space. It could have taken only a few minutes (if I was extremely lucky) or the whole 15 days (if I was unlucky).

"I think it's important to note that some of these (actually) 112 machines are quite old, and I could have done the job just as fast with 30 of the fastest workstations that we have (a DEC alphastation,which cost us little more that $10,000)," he adds. "According to some letters I got, a MasPar machine would be about twice as fast. You would get roughly the same speed as I did on a network of 40 to 50 high-end Pentium PCs."

The researcher doesn't believe the incident will have much impact on Netscape. "Everybody who understands the technical details knows perfectly well that this was doable and even easy," he says. "You have to understand what happened exactly. I did not break SSL itself. I did only break one SSL session that used the weakest algorithm available in SSL. If I want to break another session, it will cost another eight days of all my machines."

Netscape's Security Breach

A little more than a month after the French computerist performed his stunt, two graduate students at the University of California at Berkeley seriously compromised the security of Netscape's Web browser. Instead of trying to crack the key for decrypting messages using Netscape's software, the students took a hard look at the way the keys were created by the browser. The keys are supposed to be created with a random number generator. Only the students discovered the numbers generated by the software weren't so random. After making this discovery, the two students, Ian Goldberg and David Wagner, wrote a program that allowed them to guess the key to any message created by the program.

Although the software's security protocol, the Secure Socket Layer, was sound, a design and implementation error in Netscape's Navigator software made it vulnerable. The way SSL provides security is by encrypting the communications channel between the client (Navigator) and the server. The encryption is initialized by a "key," which is like a combination lock. If you can guess the combination, you're golden. Netscape made its keys very easy to crack by basing them on some easy-to-guess information, like the time of day.

Netscape quickly confirmed that the security of its software had been breached and within a week had a software fix on the Internet to cure the problem. It admitted that the students had found an Achilles' heel in the software, but one that that could be remedied quickly. "We plan to address the problem by significantly increasing the amount of random information that cannot be discovered by external sources from approximately 30 bits to 300 bits," the company said in a statement. "In addition, the random information will be made more difficult to replicate because we will greatly expand the techniques and sources used to generate the random information."

One problem with SSL from its inception was that it was never given adequate peer review, a criticism also leveled against the infamous Clipper chip. Netscape learned its lesson in this regard. Before sending its fix for the key-generation problem to its customers, it had it reviewed by an external group of world-class security experts.

The World Wide Web's Virtual Library listing on cryptography

http://draco.centerline.com:8080/~franl/crypto.html

5 DIGITAL CASH

Most of us have forgotten the often tedious content of the second-level economics course, "Money and Banking." At the time, it seemed hopelessly irrelevant. After all, what hope did the average college student have of influencing the Federal Reserve at its next interest-rate session?

Suddenly, with the onset of digital cash, the issues discussed in that class become quite compelling. The future of our economy is intricately hinged to the way digital cash is implemented, and the companies described in this book are at the cutting edge of this frontier.

Money and Banking and the Power of Cash

Some of the issues that we'll have to address as a society:

- The government controls our currency, the money supply, check clearing and, by extension, inflation and interest rates. What if this power falls into private hands?

- What happens to other stores of value, like gold, bonds, and collectibles?

- What happens to exchange rates and the balance of trade if a universal currency takes over?

- Does our economy become less stable with the introduction of a new currency? Or does the private, nonpolitical nature of the proposed systems ensure more stability?

The Internet has made exchanging information and data easier than ever. Electronic mail and documents zoom between remote sites at the speed of light. Millions of users find new friends and business associates every day. They trade the latest poop on their various and sundry interests with people who share those interests, people they never would have found without the myriad links of the Net. But when it comes to the exchange of goods and services, barriers exist that commercial interests are just beginning to batter.

A collection of links and pointers to payment schemes for the Internet

http://ganges.cs.tcd.ie/
mepierce/Project/
oninternet.html

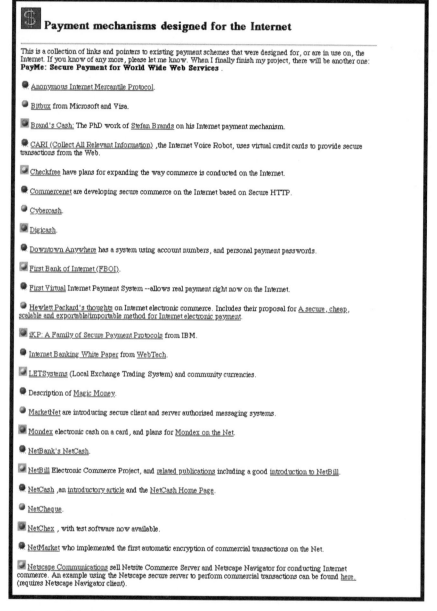

The mall has been built. But there are very few sellers, and fewer buyers. And the reason is that shopping is too hard. The question is, will an effortless transfer mechanism make people buy? And even better, will an effortless collection mechanism encourage people to sell? This isn't as trivial as it sounds. The invention of the Touch-Tone phone spawned voice mail, automated ordering, and instant banking. A small change in the dynamic can push open the door and bring thousands of new players into an industry, which will then attract customers.

Consider the impact that Federal Express had on the catalog business. Before FedEx, buying something by mail meant uncertainty. It could take days or weeks before a package was delivered. The seemingly innocuous introduction of overnight shipping changed that. Gone was "allow three weeks for shipping and handling," and in its place was the extraordinary service of, for example, MacConnection. Call (800) 800-1111 at 2:30 in the morning. Order a computer or a modem or some software. For $3, your order will arrive on your desk by noon the same day. Without sales tax. Without waiting in line. Suddenly, mail order is better than retail in every way. Simply because of overnight delivery.

The question is, will an effortless transfer mechanism make people buy? And even better, will an effortless collection mechanism encourage people to sell?

FedEx Home Page

http://www.fedex.com/

A major barrier to commerce on the Internet is how to pay for the goods and services offered there. The easiest way to pay for anything—as anyone saddled with the burden of credit card debt can attest—is with plastic. So it shouldn't come as any surprise that initial Internet payment schemes have tried to exploit the positives of plastic.

These positives haven't been overlooked by commercial online services. Both Prodigy and America Online have attracted vendors who

> **The invention of the Touch-Tone phone spawned voice mail, automated ordering, and instant banking. A small change in the dynamic can push open the door and bring thousands of new players into an industry, which will then attract customers.**

sell their wares online and accept credit card payments. But these efforts have had mixed results. Yet any indications that there may be some consumer resistance to the idea of buying things online haven't deterred commercial interests, who have taken America's contribution to Western civilization—the mall—and transplanted it to the Infobahn.

The Virtual Mall

With a few exceptions, all the electronic malls built to date are stopgaps. They use a poorly integrated hodgepodge of technologies and providers to simulate the shopping experience of a mall.

An early example was computer shopping via Prodigy. A wide array of merchants appeared during Prodigy's early years, including Sears, the U.S. Postal Service, and other national brands.

In a nutshell, the Prodigy experiment was a terrible failure. On a good day, the 2 million members of Prodigy were purchasing as little as $50,000 worth of goods—less than the total revenue of one good Sears store.

Why didn't it work? A few reasons. First, delivery took too long. The merchants weren't really wired, so after an order was placed, it was printed out and handed to an operator, who rekeyed it into a telemarketing system. Sometimes orders collected for a week or more before they were entered. Ordering stamps via the postal service's toll-free number brought them in as little as three days. Ordering them through Prodigy took up to a month.

Second, selection was limited. Unlike a full-color Lands' End catalog or the bounty available at a Price Club, the choices available online were few and far between.

Third, shopping online cost money. Toll-free numbers are just that: free. Prodigy charges a monthly fee, requires a local call, takes longer, and ties up the telephone and the computer.

Fourth, the merchant gave e-shoppers no advantages. There were no special sales, no discounts for participating in the automation, no extra perks or bonuses. "Why bother?" was the response from most people online.

Finally, the merchants did nothing to adopt the one-to-one approach. Instead, they were focused on forcing the retail/telemarketing model into a new medium. If you bought size 32 underwear this week and went back to buy more next week, the computer had no memory at all of who you were and what you wanted. There was no easy way to do all your anniversary and birthday shopping ahead of time. There was no synergy across merchants—the postal service didn't offer special envelopes to people it discovered were buying record albums from Sam Goody.

In short, a valiant effort but one that was doomed to failure because it ignored the consumer and focused on the needs of the merchant. The single greatest success on Prodigy points to the failures of its mallmates.

PC Flowers is a tiny FTD-authorized florist with eight employees and a store that receives little walk-in business. But for nearly a decade, they've had a storefront on Prodigy. They make it easy to order most of the FTD catalog, and their flowers are delivered within a day, at the same price as from any other florist.

The success of PC Flowers is legendary. Within two years, they were the single largest user of the FTD system. They have greater sales per employee than any other florist in the United States.

By offering a combination of convenience, quality, speed, and fair pricing, PC Flowers demonstrated that online shopping can work.

> **On a good day, the 2 million members of Prodigy were purchasing as little as $50,000 worth of goods—less than the total revenue of one good Sears store.**

Presenting Digital Cash

The Home Page For PC Flowers & Gifts–a pioneer in online shopping

http://www.pcgifts.ibm.com/

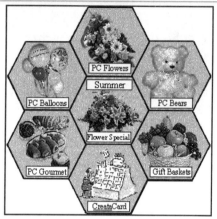

Gift Reminder

Choose one of these holidays or enter in an occasion of your own choice along with your email address, and PC Flowers & Gifts will remind you one week in advance of any of the following major holidays:

Holidays

    Apr 26 Secretaries' Day
    May 14 Mother's Day
    May 29 Memorial Day
    Jun 18 Father's Day
    Jul 4 Independence Day

Custom Event

Month/Day [ ]/[ ]

Email Address [   ]

[Submit] [Reset]

Frequently Asked Questions

- Click here to find out why more people have shopped PC Flowers & Gifts than any other interactive service in the world over the past six years.

- Click here for more information on roses.

- Click here for more information on plants and flowers.

<u>Customer Inquiry Form</u>

Since the dark, early days at Prodigy, the industry has regrouped and is now preparing to try again. Every online service is trumpeting future revenues from advertising and shopping, and special interest malls are springing up as well. Most have learned a bit from past failures, but no one has put together the right combination of benefits to attract a horde of consumers.

AT&T has become a mall developer through its PersonaLink Services network. The service, which went online in September 1994, costs $9.95 per month. It can be accessed by phone line or by wireless connection through Ardis, a wireless data carrier. It is also tied into SkyTel, a wireless paging service.

AT&T is playing to its strength. This mall is available in a multitude of ways, making it easy for people to get there. But to date, it hasn't attracted the combination of ease, discounts, and merchants that will make it work.

CompuServe has been in the mall business longer than any other existing service, including Prodigy. The CompuServe mall is a graphically disappointing place, with splash and color coming from the service's monthly magazine. The deal with the mall is simple: It's free for CompuServe members to visit. It can collect credit cards and orders and makes them easily available to merchants in a downloadable text file.

Some businesses have had a long, profitable run on Prodigy. The keys to their success are

- a small but loyal customer base

- the ability to promote themselves through the CompuServe magazine

- special discounts and promotions available online only.

Larger companies that haven't dedicated time to focusing on this niche haven't fared nearly as well as smaller companies that learned to work the system and focus on their customers.

Users with devices using General Magic's Magic Cap operating system—such as Envoy, made by Motorola, and Magic Link, made by Sony—can use a whole new technology to shop. The company's Telescript intelligent agent software can dispatch "personal assistants," which are something like electronic genies, into the network to do the user's bidding. These cyberjinn can filter incoming messages and news clips, grab information for the user on the basis of criteria he or she designs, and perform online transactions, such as completing purchases on eShop's electronic mall, called Market Square Mall. Merchants who have storefronts on Market Square include Hallmark Cards, Tower Records, Lands' End, Lexus/Nexus and, Premium Advantage, a travel services company based in Omaha, Nebraska.

EShop, of San Mateo, California, designed the system that makes it all work. It also provides storefront design consulting services to Market Square retailers.

The goal of the system is to let the agent do the tedious work of shopping around. When fully executed, the system will query the user as to what's needed and what the user's budget is. The agent can then shop around—spending hours if necessary—to collect a market basket of options. The user is presented with the options and can quickly choose the best one for her needs.

How does this help a small company? Premium Advantage, which used eShop to design its online presence, is very optimistic about its business prospects on the electronic mall. According to what it terms conservative estimates, it expects to quadruple its revenue during the next two years using PersonaLink as a merchandising avenue.

Initially, Premium Advantage's storefront will graphically display the company's services: worldwide hotel bookings at discount rates, transportation arrangements, and the acquisition of tickets to sporting and entertainment events. At the storefront, users can sign up for membership in Premium Advantage. Members can use PersonaLink's messaging service to ask Premium Advantage to appoint a personal concierge to fulfill their service requests.

Future plans by Premium Advantage include allowing users to book their own travel arrangements and to take virtual tours of hotels they're considering staying in.

In the one-to-one future, Premium Advantage can take advantage of the habits, preferences, and needs of its members by offering them exactly what they want, exactly when they want it. For example, hotels are always struggling with yield management. An empty room is money down the drain. If Premium Advantage is able to create a community of travelers, and it knows when they want to travel, it can negotiate for this last-minute space at significant discounts.

If Premium Advantage knows that it has 200 people willing to travel to Hong Kong for a week-long vacation, it can find them rooms, plane seats, etc., and deliver them (knowing they'll be paid) directly to its members.

Unlike these proprietary malls, which are off the Internet, marketplaceMCI, which went online in April 1995, is on the World Wide Web. Combined with MCI's efforts to bring Fox and its Delphi online service to the masses, marketplaceMCI is a strategic effort to combine MCI's proven expertise in telecommunications with the often elusive goal of capturing consumers' desire for information and entertainment.

MarketplaceMCI is trying to get consumers to buy over the Internet

One of the flagship sites on marketplaceMCI is @Bat, the official window on Major League Baseball. The site is chock full of statistics, news, photos, and trivia about baseball. That alone would provide baseball with a much-needed PR shot in the arm.

But the site is designed to generate a profit as well. Easily available at the click of a button, the MLB Clubhouse Store offers a wide range of merchandise, some of which isn't available in any store. MCI and MLB hope that the combination of secure transactions and unique merchandise will be sufficient incentive to get users to loosen their purse strings.

Major League Baseball on marketplaceMCI has statistics, news and merchandise

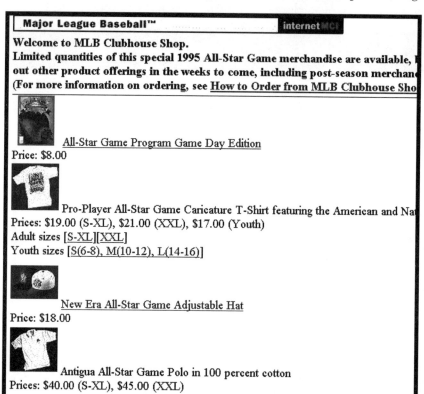

Some retailers that can be found on marketplaceMCI include Art Access, a seller of fine art celebrating nature; Covey Leadership Center, a training firm based on *The Seven Habits of Highly Effective People*; Damark, a mail-order discount computer and electronics retailer; Dun & Bradstreet Information Services; Hammacher Schlemmer & Co., an upscale mail-order retailer; Intercontinental Florist, a florist offering worldwide delivery of flowers; Mac Zone and PC Zone, a mail-order computer products store; OfficeMax, an office products store; and Shocker T-shirts and Apparel, whose name speaks for itself.

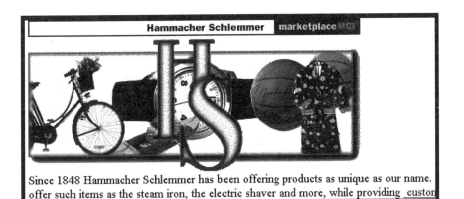

Hammacher Schlemmer is one of the retailers on marketplaceMCI

While this is an impressive start, it doesn't approach the completeness and integration a virtual shopper demands. To make up for the tactile pleasure of touching items, trying them on, and taking them away, marketplaceMCI will have to offer a selection bigger than any mall, together with back-end smarts that allow the mall to learn about the user and respond.

From the point of view of developing Internet commerce, the most important feature of the MCI Web mall is its approach to securing transactions at the site. Preventing the poaching of sensitive information on the Internet is one of the biggest challenges facing commercial developers of the Net. Hackers have penetrated the data of Internet service providers in the past and, to the horror of these providers, thousands of credit card numbers have been stolen. Cases like these have made potential Net vendors cautious.

MCI has responded to this need by developing special software. While creating client software solves a number of technical problems (as we'll see throughout this book) it creates a host of marketing problems that may be difficult to overcome.

While anyone can browse at marketplaceMCI, a consumer needs MCI's special client software, called InternetMCI Navigator, for encrypting sales information. One problem with this approach is that when a potential customer visits the mall for the first time and tries to make an impulse buy (and how much mall shopping has to do with impulse buying?), he can be blocked from exercising his shopping instincts by not having MCI's special software. This is similar to having to get a special charge card to shop at a given store. Once you've got the card, shopping is pretty easy. But many consumers won't overcome the first hurdle and sign up. MCI attempts to address this problem by

> **To make up for the tactile pleasure of touching items, trying them on, and taking them away, marketplaceMCI will have to offer a selection bigger than any mall, together with back-end smarts that allow the mall to learn about the user and respond.**

offering the software to users for free. Also, if a user tries to buy something without the client software, a help screen pops up informing the customer where he can download the program.

Making the Net Safe for Commerce

MCI isn't alone in developing a method to protect business transactions on the Internet. Other players include Netscape Communications, International Business Machines (IBM), and credit card companies Visa and MasterCard.

So far, most efforts have been directed at creating digital credit, not digital cash. This is a logical place to start, since the collection mechanism for credit cards is already in place. But it doesn't address a few critical issues. The size of transactions is limited to mid-size expenses—nothing too big or too small. Also, it's hard to get money out of the system and to become a merchant.

But digital credit is an important first step. It will establish patterns in which producers get used to selling things and consumers get used to buying them. It will point out the benefits of digital cash and make it clear that true digital cash is worth the effort.

Here's an overview of some of the key players in the first generation of digital credit.

Netscape, which went public in 1995 in one of the more celebrated initial public offerings of the year, makes one of the most widely used Web browsers, the software used to view graphic pages on the World Wide Web, where commercial activity on the Internet is hottest. Netscape now includes in its browsers an encryption system that creates a secured channel to prevent unauthorized personnel from tapping into a network, a method of authenticating the identities of parties entering a transaction, and measures to protect the integrity of messages traveling over a network and ensure they can't be altered en route to their destinations.

The Home Page for One of the popular browsers for the Web: Netscape

http://home.netscape.com/

Netscape's News and Reference page

http://home.mcom.com/newsref/index.html

Netscape, along with America Online, CompuServe, IBM, and Prodigy, has also invested in Terisa Systems, a joint venture of Enterprise Integration Technologies, a developer of electronic commerce systems, and RSA Data Security, a major developer of data encryption systems. The participation of these online heavyweights in Terisa has been hailed by industry insiders and observers as a breakthrough for Internet security because it may avert a battle over standards for protecting transactions on the net.

The Home Page For RSA Data Security

http://www.rsa.com/

Terisa has developed a technology called Secure Hypertext Transfer Protocol (S-HTTP). One of Terisa's backers, CompuServe, contends that S-HTTP is so good it rivals the online service's own security. That's why CompuServe plans to expand its Electronic Mall, which went online eleven years ago and has more than 170 merchants, to the Web. Web server makers have already begun to incorporate Terisa's protocol into their products. Open Market, of Cambridge, is marketing a product called Secure WebServer that incorporates S-HTTP.

Although IBM has also signed on to the Terisa venture, it has separately developed a method, called iKP, for making secure multiparty transactions—transactions, for example, involving a consumer, a merchant, and a financial institution.

The IBM scheme works like this:

A buyer submits his purchase order and encrypted credit card number to a merchant. The merchant is able to forward the pricing information and credit card number to the appropriate financial institution for credit approval without having to go offline.

The beauty of iKP is that the consumer's credit card number remains encrypted while in the hands of the merchant. Only the financial institution has the capability to decrypt the number and look at it. This reduces the risks of credit card fraud by the merchant or by hackers who might break into the merchant's computer system. With iKP, IBM boasts, credit card transactions conducted on the Internet will be safer than those executed in traditional retail channels.

A big customer for iKP will be Europay, which has issued 57 million credit and debit cards. Europay will incorporate iKP into "smart cards." Smart cards are a little thicker than credit cards and contain a computer chip, which gives them a degree of intelligence. Europay's iKP smart cards will be used in conjunction with card readers attached to televisions, telephones, or PCs to perform a number of transactions.

IBM page mentioning Terisa and iKP

http://www.ibm.com/Features/InetWorld95/pr4.html

The proliferation of smart cards is a crucial development for the acceptance of digital cash, since most e-cash schemes call for the cards to have some role in them.

Eurobank, Visa, and MasterCard support a standard for smart cards named after the initial letters of the three firms: EMV. EMV-based smart cards rely on encryption technology to authenticate the user, keep data confidential, and prevent data tampering. They're much smarter and more secure than the magnetic-strip cards to which most Americans are accustomed. Since the cards contain a chip (essentially they're a hand-held computer without a keyboard), they can receive and process information, as well as store data such as credit or debit card numbers, electronic checks, health-care information, frequent-flyer miles—you name it.

One advantage of smart cards over magnetic-strip cards has induced a powerful business segment to support the devices. Banks like the idea of the cards, which have gained greater acceptance in Europe than in the United States, because they can reduce transaction processing costs. Unlike ATM (automated teller machine) and POS (point-of-sale) devices that require transactions to be authenticated at a remote location, smart-card transactions can be verified locally at a merchant's smart card terminal. That cuts the cost of those transactions.

It also opens the door for electronic cash. If transactions can be confirmed locally, then the next step is to allow "cash" to be stored on the cards.

- EPS Smart Card Enterprises plans a pilot program with 50,000 cardholders in Delaware with McDonald's and Giant Food.

- Chemical Bank is testing e-cash for its employees to use in the company cafeteria.

- MasterCard has plans for an e-cash experiment in Canberra, Australia.

- Visa will likely roll out smart-card-based e-cash at the Summer Olympics in Atlanta in 1996.

The ultimate goal of e-cash is to create a multicurrency system that not only allows retailers to accept cyberbucks from a smart card, but also allows cardholders to exchange digital money with each other as well as over a network.

The Monetary Revolution

These experiments—and others that we will discuss later—have the potential to change the way we look at money. Their basis, digital money, has the power to change our financial lives and shake the foundations of the global financial system.

Cyberbucks are the ultimate—and, many observers feel, the inevitable—exchange for a wired world. With digital simoleons there will no longer be the need to fumble for change. Just load up your smart card with digital currency and pay for everything off the card.

Remember, though, that digital credit is largely a one-way street. Unlike digital cash, digital credit makes it easy to spend money, but still requires a challenge to all small merchants to collect money. And under the new paradigm, everyone will have his own lemonade stand, her own way to collect money.

Electronic cash will let you shop online, zap money to a friend on the Internet, or pay for goods and services through your cable-enabled TV. All your current cash transactions—movies, restaurants, taxis, and more—could be paid for in e-cash. And because the depository of the new currency is programmable, businesses could tailor spending from the cards to specific accounts. The petty cash card could be limited to purchases at a certain maximum or to specific categories of goods, like copying services or office supplies.

In a paper titled "Digital Cash and Monetary Freedom," Jon W. Matonis, founder of the Institute for Monetary Freedom, outlines the elements of a true digital cash system, in contrast to the secured credit schemes we looked at earlier.

The Full Text of "Digital Cash and Monetary Freedom" by Jon Matonis

http://www.isoc.org/HMP/PAPER/136/html/paper.html

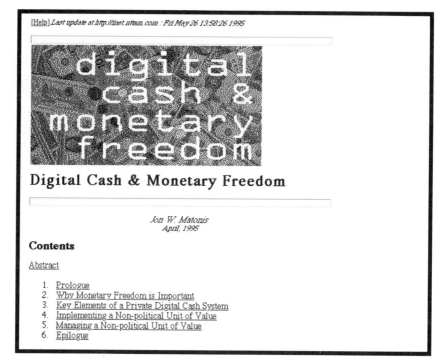

"The transition to a privately-operated digital cash system will require a period of brand-name recognition and long term trust," writes Matonis. That may be why banks have more of an upper hand in this arena than other institutions. "Some firms may at first have an advantage over lesser-known name-brands, but that will soon be overcome if the early leaders fall victim to monetary instability. It may be that the smaller firms can devise a unit of value that will enjoy wide acceptance and stability (or appreciation).

"True digital cash as an enabling mechanism for electronic commerce depends upon the marriage of economics and cryptography. Independent academic advancement in either discipline alone will not facilitate what is needed for electronic commerce to flourish. There must be a synergy between the field of economics which emphasizes that the market will dictate the best monetary unit of value and cryptography which enhances individual privacy and security to the point of choosing between several monetary providers. It is money, the lifeblood of an economy, that ultimately symbolizes what commercial structure we operate within."

Matonis argues that what is about to happen in the world of money is nothing less than the birth of a new Knowledge Age industry: the development, issuance, and management of private currencies. "Once seeded," he says, "digital cash as the representation of binary value will pave the way to a further off-network revolution in money.

"The opportunity to launch an alternative monetary system on a grand scale simply has not been available until recently," he writes. "But only lately with a global, inter-networked society can we truly say that the established monetary order is susceptible to challenge. Specifically, the Internet provides (1) ease of mass issuance and circulation, (2) accessible encryption technology, (3) affordable currency transferrable infrastructure, and (4) real-time conversion between competing units. Essentially, for the first time ever, each individual has the power to create a new value standard with an immediate worldwide audience. This should serve as a friendly warning to the clearing associations, banks, and financial service providers of the current paradigm."

What digital cash provides the world is monetary freedom, the monetary freedom to preserve the free-market economy. That economy depends on a flexible, responsive monetary system, which needs unbridled competition to work. The underlying philosophy of digital cash is that that competition should extend to currency. The market for cash should be no different from the market for mouthwash and hair spray.

"When a single currency issuer, such as the 'Fed,' controls the supply of money and the specific units being transacted, the potential exists for monetary manipulation and an overbearing control of the economy," Matonis writes. "With the unprecedented growth of the Internet, the standards for electronic commerce are still evolving. Neither the US dollar, nor any other governmental unit, has gained a foothold into this economy. The monetary landscape is ripe and wide open and private currencies should infiltrate now."

Matonis goes on to say: "If all digital cash permits is the ability to trade and store dollars, francs, marks, yen and other governmental units of account, then we have not come very far. Even the major card associations, such as Visa and MasterCard, are limited to clearing and settling governmental units of account. For in an age of inflation and government ineptness, the value of what is being transacted and saved can be seriously devalued. Who wants a hard drive full of worthless digital cash? True, this can happen in a privately-managed cash system, but at least then it is determined by the market and individuals have choices between multiple providers."

> **Matonis argues that what is about to happen in the world of money is nothing less than the birth of a new Knowledge Age industry: the development, issuance, and management of private currencies.**

> **The underlying philosophy of digital cash is that competition should extend to currency.**

According to Matonis, a private digital cash system must be

- Secure—The transaction protocol must ensure a high level of security. A person should be able to pass e-cash to another person without a third person altering or reproducing the transaction.

- Anonymous—Transactions conducted with paper money are anonymous. If digital cash is to gain acceptance as common tender, it must be anonymous, too. Governments won't like this idea, because anonymity is the characteristic of digital cash that will make their currencies irrelevant.

- Portable—Digital currency, like its paper counterpart, must be usable anywhere. You should be able to use it online or off. In fact, digital cash should have greater portability than paper money because it will be able to travel through cyberspace, a trick old-fashioned currency is unable to duplicate.

- Two-way—Digital cash must be able to pass between peers, not just between consumer and merchant, as credit-based systems operate today. You should be able to download your children's allowances into their digital wallets or pay off your debts to your friends at the end of a night of poker.

- Offline—If two people want to exchange cyberbucks, they shouldn't have to go online to authenticate the transaction, as merchants have to do to authenticate a credit card purchase.

- Divisible—For every dollar there are cents, and that should be true for digital cash, too. The e-cash should be fungible so that reasonable portions of change can be made. The smaller these change units are, the better for facilitating high volumes of small-value transactions.

- Infinitely durable—Digital cash should not expire. It should maintain its value until it is lost or destroyed unless the issuer of the money debases its value or goes out of business.

- Widely acceptable—If my Net lettuce is only recognized on my block, it isn't going to be of much value to me. True digital cash should have wide acceptance.

- User friendly—The e-cash should be simple to use from both the spending perspective and the receiving perspective. Simplicity leads to mass use, and mass use leads to wide acceptability. Pulling a sawbuck from your wallet and handing it over to a pal is mechanically, if not emotionally, easy. A similar transaction using digital cash should be as simple.

- Free of unit of value—The value of digital cash should be driven by market forces, not government ones.

Who will be the new providers of digital cash? "Opportunities abound for almost anyone, but in reality the greatest advantage currently goes to the online shopping malls and the large merchant sites on the Internet, such as Open Market, Internet Shopping Network, and Net Market," Matonis writes. "For it is this group that will directly influence the payment channel between consumer and merchant through their extensive contact with both."

He continues: "This influence can be utilized to their advantage to build preference for their site through money issuance in much the same way that various forms of scrip and coupons build customer loyalty and guarantee repeat visits."

Other unit providers identified by Matonis are Internet service providers, bulletin board system operators, content publishers, card-based payment networks, and well-known manufacturing or service companies. "They all share in common the existence of an extensive base of online customers," he notes. "As the new digital cash providers, national brand names, such as Coca-Cola, Microsoft, and IBM, find themselves in an enviable position to capitalize on their global name recognition."

But before Coke moola takes over the world, let's look at some of the players in the digital cash game today.

6 MONDEX

In the English town of Swindon, customers at the local McDonald's buy Big Macs without touching a bank note; pub crawlers at Bass Taverns keep the taps running without tapping their wallets; and grocery shoppers pay for their provisions without currency changing hands. Citizens of Swindon, you see, are participating in a pilot project testing Mondex, a smart card for dispensing digital cash.

How It Began

Mondex is the cerebral offspring of two executives at National Westminster Bank (NatWest) in London, Tim Jones and Graham Higgins.

Jones, who graduated from Cambridge University in 1976, began his career working for Shell UK Oil and Lucas CAV. After a stint playing with a rock band in Brighton, he joined NatWest's Operational Research Group in 1983. He worked on point-of-sale systems and debit cards until

1988, when he was appointed senior executive of the bank's Card Strategy Group with the task of guiding the bank through the turbulent period of Visa/MasterCard duality, the breakup of the Joint Credit Card Company, and the internationalization of the bank's debit card, called SWITCH. Jones has also been active in pan-European affairs. He is a director of Eurocheque International and was involved in the formation of Europay as well as the development of the EDC/MAESTRO cross-border debit product.

Before Higgins joined Jones in developing Mondex, he was a member of NatWest's Card Strategy Group and was involved in the launch of the bank's Visa program and retailer card services as well as undertaking extensive research into smart-card-related technologies.

In March 1990, Jones, then deputy director of payment services at NatWest, and Higgins, then a manager in the Card Strategies Group, cooked up the idea for Mondex. By April 12, the company had applied for its first patents.

During 1991 and early 1992, the Mondex team began establishing relationships with the major electronics companies it would need to develop its digital cash system—Dai Nippon, Hitachi, Panasonic, and Oki—and engaging in market research. Forty-seven consumer focus groups in five countries—the United States, France, Germany, Japan, and the United Kingdom—were gauged for attitudes toward electronic cash. Detailed market research was carried out with more than 5,500 consumers altogether.

The founders of Mondex

In 1992, British Telecom was enlisted to work on the introduction of Mondex into the UK. And the Bank of England was informed that Mondex and NatWest intended to establish Mondex as a global electronic cash system. In March of that year, Mondex launched a trial program at NatWest's London office. More than 6,000 employees were given smart cards enabling them to pay for goods and services at the staff restaurants and shops. By the end of 1994, 1 million purchases had been made using the technology.

Midland Bank agreed to be Mondex's second banking partner in the UK in 1993. By the end of the year, NatWest and Midland announced they would be 50-50 partners in Mondex UK, the company running the system in Great Britain, and that Mondex was going global.

In 1994, Mondex began approaching major financial institutions worldwide and asking them to join in a new company, Mondex International, capable of becoming a global payment system. By April of that year, product development specifications were released enabling manufacturers to begin developing Mondex-compatible products. By the end of the year, more than 120 manufacturers had expressed interest in developing Mondex devices. In September, Mondex UK announced that more than forty major High Street retailers had agreed to accept payment with Mondex e-cash when it was rolled out nationally.

Meanwhile, Mondex International landed two major players in the Far East and North America. In October 1994, HongkongBank agreed to be a Mondex franchisee serving Hong Kong, China, India, Indonesia, Macau, the Philippines, Singapore, Sri Lanka, Taiwan, and Thailand. And in May 1995, two of Canada's largest banks—the Royal Bank of Canada and the Canadian Imperial Bank of Commerce—were welcomed into the Mondex fold.

HongkongBank chairman John Gray declared Mondex "a significant development in modern banking" and one that would bring wide benefits to the customers of his bank.

The Canadian banks said that they would launch a pilot program for the system in 1996 and have a national rollout in 1997. "Mondex will revolutionize the way people pay for goods and services," observes R. M. (Bob) Juneau, Royal Bank's senior vice president for card and point-of-sale services. "Mondex can be used for all sorts of payments, from everyday low-value transactions to major consumer durable purchases. And just like cash, the Mondex system allows users to make cash transactions person to person.

"Whatever amount of ordinary cash you normally carry around with you in your wallet or purse, you can have an equivalent amount of electronic cash stored in your card," he adds.

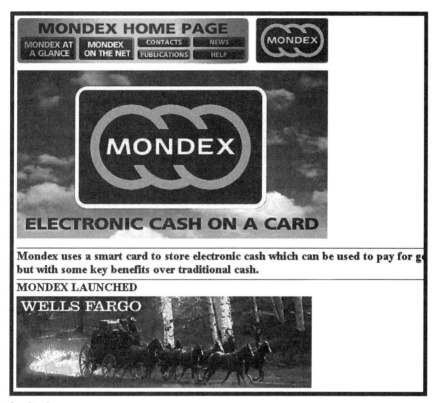

Mondex Home Page

http://www.mondex.com/mondex/home.htm

Swindon

Mondex UK's pilot e-cash project began in July 1995 in Swindon. National Westminster and Midland picked Swindon for their pilot because it's the Columbus, Ohio, of the United Kingdom. It's a perfect demographic microcosm of the country. It also has excellent communications links and is in the heart of Great Britain's high-tech heartland.

Swindon–Mondex's Test City in England

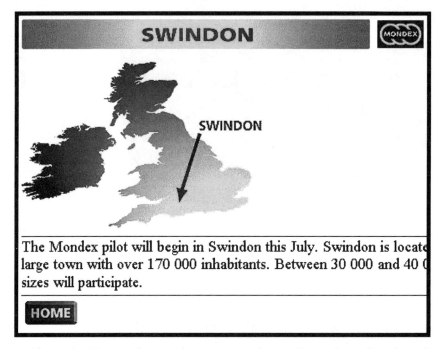

About 1,000 Mondex cards were issued initially in Swindon, but the banks expect to distribute up to 40,000 cards as the pilot picks up steam. Everyone—including children as young as five years old—who lives, works, or shops in the area can apply for a card, regardless of whether they're customers of National Westminster or Midland.

The banks also plan to recruit about 1,000 retailers into the program. As early as a month before the Mondex card was rolled out in the area, according to one eyewitness, retailers, pumped up over the pilot, had their terminals installed and on display on their premises.

Dai Nippon makes the cards being issued for the Swindon pilot. The cards use Hitachi EEPROMs and conform with ISO 7816 cards, the international standard for IC cards. The chip inside the card has been programmed to act as an electronic "purse." The purse can be loaded with value, sometimes referred to as tokens, until it is used as payment for goods or services at retailers participating in the program. The card can be "locked" by pressing a button on it and unlocked through the use of a personal identification number (PIN) chosen by the card owner.

Of course, it's possible to build much more into these cards, but Mondex is shying away from multifunctionality. Getting people to accept as abstract an idea as digital dough will be hard enough for

Mondex without confusing the issue with more bells and whistles.

Consumers participating in the Swindon e-cash pilot receive a Mondex card and a balance reader that's about the size of a key fob. A user can load the card into the reader to obtain the balance on the card. For the first six months of the pilot, participation is free of charge, but after that time consumers will have to pay £1.50 a month. That shouldn't be much of a problem, if Mondex's market research is correct. When Mondex tested the waters for consumer reaction to smart-card cash, 94 percent of the people polled liked the idea and 75 percent of the survey sample said they wanted a digital cash card even if they had to pay for it. "It's probably the most positive financial services research that any of us have ever seen," Ann Adams, a senior manager with Mondex, told *World Card Technology* magazine.

When Mondex tested the waters for consumer reaction to smart-card cash, 94 percent of the people polled liked the idea and 75 percent of the survey sample said they wanted a digital cash card even if they had to pay for it.

A Mondex "Smart" Card

WHAT IS MONDEX?

The Mondex card is an integrated circuit (IC) card; a "smart" card - in it.

The card will take the form of ISO 7816 IC cards - the international

This microcomputer has been programmed to function as an "electr

The electronic purse can be loaded with "value", which is stored un outlets that participate. The current balance on the card can be chec

The electronic purse can also be locked using a personal code so th

HOME

Consumers can also obtain an optional electronic "wallet," which includes two extra cards. The wallet is a pocket-sized device with a keyboard and a screen. The wallet lets a person keeps a separate store of value, say at home or in a hotel room, while maintaining a minimum balance on his or her card. This gives the user some added security; if the card is lost, only the minimum balance will be gone. The wallet gives a cardholder the ability to check the last ten card transactions. Transfers of value between individuals can be done through the wallet. Families, for example, can transfer value between cards. Children could have their own cards, and parents could transfer "pocket money" or a weekly allowance to the cards. Mobile retailers, such as taxi drivers, could also use the wallet to collect compensation for their services. The wallet and spare card package will also be free in Swindon for six months. After that the charge will be £3.50.

Although Mondex's digital cash wallets are barely off the drawing board, the company is already thinking of tinkering with the interface. It's considering incorporating an infrared protocol into the wallet so cardholders could zap cyberbucks to each other across a room. (The original prototypes of the card called for such an interface.)

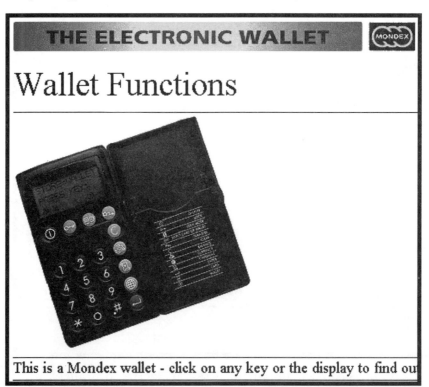

An Electronic Wallet From Mondex

E-cash can be loaded into the Mondex cards through a Midland or NatWest ATM or by means of special Mondex-compatible screenphones made by British Telecom (BT). The phone acts as a personal ATM. To download value by phone, a user inserts his card in the phone, dials a number that connects him to his account, enters a PIN number after he makes his connection, and punches in the amount of money he wants. About 1,000 phones will be distributed to selected customers in the Swindon pilot and a number of Mondex-compatible pay phones will be installed in the area.

To buy something in a store, the cardholder inserts her card into a retailer's Mondex terminal. The retailer punches in the amount to be deducted from the card. If there isn't enough value in the card to cover the purchase, the transaction is blocked; the cardholder sheepishly leaves the store, or else dials up her account and transfers some more e-cash to her card. If a buyer brings a purchase back to the retailer, a refund can be loaded directly onto the customer's Mondex card through the retail terminal.

A log of each transaction is kept at the retailer's terminal. That log includes the value of the transaction and the card's "narrative," or identification. The log can be printed at the end of the day and an audit trail created. This feature is being used by Mondex to track spending patterns by participants in the Swindon pilot.

When a merchant is ready to go home, he can transfer the day's proceeds to his own Mondex card and deposit them to his bank by phone. No more runs to the depository window with a bag full of money to attract thieves.

Payments between Mondex cards can also be made over the phone. This makes the card suitable for transactions over the Internet. If someone wants to sell something on the Net, she no longer needs to worry about being certified by Visa or MasterCard, as long as the buyer has a Mondex card.

The beauty of smart-card technology is the cards' ability to talk to each other. That makes verification of transactions very secure. Each Mondex card has an identifying narrative stored on its chip—the cardholder's name or initials, for instance. The narratives are transmitted between the parties when their Mondex phones connect so they can see who they're dealing with.

This transfer, though, isn't as secure as Mondex would like it to be. According to Jones, it is possible for a Net hacker to send someone else's legitimate narrative to Mondex while having the card transfer value to the hacker's card. When a cardholder checked her transaction record, she'd see that the value was misdirected, but it would be too late. This is

The beauty of smart-card technology is the cards' ability to talk to each other. That makes verification of transactions very secure.

> **The value on the card can be discovered by Mondex by "interrogating" its chip. The company is confident it will be able to discover deliberate tampering with a card.**

similar to someone charging items with your credit card, without the 90-day chargeback safety net that credit cards offer.

One of the purposes of the Swindon pilot is to increase the comfort level of merchants and consumers about electronic cash in general and Mondex in particular. Mondex will also be looking at buying patterns among consumers to determine what kinds of stores they're using the card in and how much they're spending per transaction.

Another aspect of the pilot project Mondex will be looking at is consumer reaction to lost or damaged cards. Mondex intends to treat the cards as money. If you lose it, you lose the value in it, just as if you lost your wallet. However, a finder might be more likely to return your smart card than your wallet, since only a cardholder can unlock his card, while cash from a wallet can be spent immediately.

Damaged cards are a different story. If a card is accidentally damaged, Mondex will restore the value to the cardholder. According to Mondex, all breakage in the program will be studied in minute detail. The value on the card can be discovered by Mondex by "interrogating" its chip. The company is confident it will be able to discover deliberate tampering with a card. In those cases, the value in the card will not be refunded to the cardholder. Concern over lost and damaged cards is one reason why Mondex is limiting the amount that can be downloaded into a card during the pilot to £500. The company also plans to launch an insurance plan to further protect cardholders should their cards be lost or damaged.

A *Mondex Terminal*

http://www.mondex.com/mondex/shop.htm

Unaccounted System

These concerns over lost and damaged cards arise because Mondex is an unaccounted system. This distinguishes Mondex from some other budding rivals in the e-cash arena, but it has also opened it up to some scalding criticism. One critic has been Banksys, the Belgian operator of an electronic purse program called the Proton Scheme. Banksys argues that Mondex's unaccounted system, by failing to audit a cardholder's expenditures, will undermine public confidence in digital cash and reduce the chances of discovering fraud in the system. Yet it's the unaccountability of the Mondex system that makes it a true digital cash plan. Without anonymity, you don't have cash. Without anonymity, you don't have privacy. As NatWest's Jones pointed out in *World Card Technology*: "I'd think there'd be considerable civil liberties issues if there were [central control]. What Mondex is trying to do is to draw a very careful line between consumers' legitimate rights to privacy and the state's legitimate right to frustrate antisocial activity such as money laundering."

Although the Mondex scheme has been characterized as an unaccounted one, there are audit safeguards built into the system. Each card

> It's the unaccountability of the Mondex system that makes it a true digital cash system. Without anonymity, you don't have cash. Without anonymity, you don't have privacy.

has a unique identifier, as does each transaction conducted through a Mondex card. This deters bandits from trying to duplicate or divert a transaction. In addition, each card contains three logs.

- A user transaction log records data on the card's last ten transactions. Those data include the value transferred in a transaction, the recipent's identifier, and the number of the transaction.

- A pending log documents the status of the current transaction. Should the transaction fail, this log can be used to determine what went wrong with the transfer.

- An exception log records unsuccessful transactions while allowing other transactions to take place. When this log fills up, the card is disabled and has to be returned to the bank that issued it to be reset or replaced.

Security

Although the Mondex system builds in unaccountability, it doesn't build in an invitation for license. The company took security issues very seriously when designing its e-cash system. It realizes the system needs a high level of security to guard it against counterfeiters, who would try to knock off the card, and hackers, who would try to mimic the card and forge electronic money, or intercept and divert legitmate cyberbuck transfers.

The layer of security most obvious to the cardholder is the lock. It can be turned on with a single keystroke and unlocked with a PIN. A card can receive value even when it's locked. That applies to a retailer's Mondex device, too, so he can lock the "till." Employees will still be able to transact business with customers, but the employee won't be able to take any money out of the machine.

After a series of incorrect attempts to enter a PIN, the card will shut down; it must then be taken to a Mondex-issuing bank, where the cardholder's identity and entitlement to the card can be verified.

The chips in the Mondex cards have been specifically designed to protect the data in them from unauthorized disclosure and modification from both physical and logical attacks. The company spent four years developing state-of-the-art protection against reverse engineering by finding the world's best reverse engineers and give them a shot at cracking the chip in their smart cards. These engineers try to work backwards from a finished chip to create the original code.

The first microchip to be used by Mondex is the Hitachi H8/310 smart card microprocessor. Into this chip, Mondex loads its security programs. These include Mondex's value transfer protocol, which uses sophisticated cryptography to protect value as it passes from one Mondex card to another. Value can only move from card to card—it doesn't reside anywhere else where it can be compromised—and it can be stored only on Mondex cards.

When a transfer is about to take place between Mondex cards, two steps, unseen by the cardholder, take place to ensure a legitimate transaction.

- Registration information is exchanged and validated by the two cards involved in the transaction.

- The card belonging to the recipient of the value sends a digital signature to the payer's card, requesting a sum. The payer's card checks the signature, debits the sum against the value stored on the payer's card, and transfers the sum to the recipient's card with its own signature attached. The recipient's card checks the payer's signature, sends an acknowledgment, with digital signature, to the recipient's card, and credits the sum to the recipient's card. This guards against duplication or unauthorized creation of value.

With hard cash, counterfeiters can get by, like horseshoe players, by being close to what they're forging. And they do that with relatively inexpensive technologies—PCs, laser printers, color scanners. The technology to reverse engineer the chip in a smart card is very expensive, almost prohibitively so for the lone-wolf forger. And even if the knock-off artist could marshal the resources to attempt a forgery, he would have to make a perfect copy. If the chip isn't perfect, the card just won't work in the system. Add to that periodic updates in cryptographic algorithms and chip design and you've created a very risky venture for a criminal.

Another headache for forgers is the rapidity with which Mondex can change its currency. With hard cash, criminals can be sure that the face on a dollar bill isn't going to change overnight. But no such surety exists for digital cash.

One way Mondex can change its currency is to reissue new cards periodically. Mondex already plans to have its cards swapped out of circulation every two years for security upgrades.

Another, and less disruptive, way to change the currency is to alter

> **With hard cash, criminals can be sure that the face on a dollar bill isn't going to change overnight. But no such surety exists for digital cash.**

the software in the cards. This can be done by software downloads, much as some online systems, like Prodigy, upgrade their software. One day a user will plug her card into an ATM or access her account by phone, and the new software will automatically be downloaded into her chit. This will allow Mondex to, in effect, change its currency in a way that will be transparent to its users and will not disturb commerce. Compare that to the chaos that would result if Uncle Sam decided to change the color of U.S. currency from green to blue.

An alternative to software downloads is the inclusion of multiple security systems in the cards when they're issued. Cards can be issued with an "A" and a "B" system, for example, with the A system turned on. To change the A system to the B system, a third card, one with a C/D system, can be issued with software instructions to, when it comes into contact with an A/B card, switch the card to the B system. If the B system includes instructions to switch A/B cards running system A to system B, a cascading effect will take effect. The A/B cards can be withdrawn from circulation during the normal replacement cycle. And when it's time to change the currency again, a E/F card can be issued to turn the switch in the C/D cards.

When Mondex designed the security for its system, it did so with an eye to doing more than just deterring fraud. It wanted to maximize the criminal's risk of being caught and minimize his opportunity to benefit from his larcenous endeavors.

With hard cash, there's no way to quickly analyze patterns in the flow of the currency. With digital cash, there is. By analyzing information from retail terminals and bank cashpoints, Mondex can analyze patterns of behavior that may indicate fraud in progress—for example, duplicate transaction numbers appearing in the system, or a consumer's card consistently redeeming amounts that would be more typical of a merchant.

As the chips in smart cards get faster, companies like Mondex will be able to use longer key lengths, making the cryptography in them tougher to break. Just as smart-card chips are getting faster, the track width of the silicon in the chips is getting narrower, which makes them tougher to reverse engineer.

"The security of the Mondex system depends on the Mondex organization maintaining the integrity of the chips which it uses and ensuring the continued soundness of the electronic architecture of the system," writes Mondex senior security manager Brian Pugh in *World Card Technology*. "Mondex security is based on continuous, ongoing refinement to achieve further levels of defense against potential future attack.

As chip cards become more sophisticated—able to carry and transfer more information—the Mondex approach to fraud containment can become ever more powerful."

Pugh continues: "The guiding spirit of the architecture of Mondex has been that the system should be flexible enough to incorporate the security enhancements identified by the banks operating the scheme in a given territory at a given time. Safety measures can be rapidly enhanced to counter potential attacks made on it in each country and in each environment. The Mondex approach is to encourage regional banking experience to reinforce overall Mondex security."

Bank's-Eye View of Mondex

In broad terms, the Mondex system resembles the cash system. Instead of a central bank, the Mondex system has an "originator" that issues Mondex value to members, who redistribute the value to retailers and consumers through the system's smart cards. Mondex International is responsible for maintaining the integrity of the product and enforcing business rules and standards.

Banks joining the system can buy a franchise and become an originator of Mondex value. Financial institutions that are franchisees also have the option of buying a share of Mondex International. Each territory in the system will have one franchise.

Some countries interested in Mondex may have their central banks become originators. In other countries, consortia of commercial banks may take on that role.

Originators are the only entities that can create or destroy value in a territory. Each originator has a "master value" card on which the manufacture of value is carried out, but Mondex can control the creation of the value on the master value cards.

Future of Mondex

Mondex officials have been cautious about predicting the demise of hard cash. They see digital money as an alternative to cash, another option among many options for consumers. Those options include credit and debit cards, although debit card growth may be dulled by the rise of electronic cash.

Mondex estimates that e-cash will carve out 30 percent of the payments market. That kind of market will attract competition, but Mondex believes it can hold on to from 63 to 83 percent of the market, with "accounted" purses, such as Visa's, holding the rest.

Banks need not be faced with a stark choice between Mondex and an accounted system, however. Mondex is looking into possible participation in accounted-purse trials by Visa, MasterCard, and Europay, even though the architectures of accounted and unaccounted systems differ radically, and even though Mondex discarded the accounted approach early on because the company felt the approach didn't make business sense.

However, cyberbucks will have a significant impact on how banks do business. Branch offices and ATMs will no longer be the only places consumers can get cash. If cash transactions can be completed over the phone, the need for expensive branches will decrease, which will make bottom-line watchers at financial institutions smile. In addition, the costs of handling hard cash could plummet. In the United Kingdom alone, this cost has been estimated to be around £2 billion ($3 billion) a year. What Mondex does to the cost of banking could very well be more significant than the company's own profits.

Of course, there are costs related to all kinds of transactions. There's an expense in posting a credit card or debit transaction. There's an expensive central audit trail behind those transactions. This is where Mondex's system, without central auditing, shines. It makes the marginal cost of each Mondex transaction "vanishingly small," according to Mondex CEO Jones.

Five years down the road, Mondex predicts it will be the undisputed global leader in electronic cash—a prediction that no doubt raises eyebrows in the boardrooms of Visa and MasterCard.

But Mondex doesn't see itself exclusively as the holder of the electronic purse. It wants to be more than just a leader in low-value transactions. The company's system was designed to be compatible with electronic data interchange (EDI) so it has the capability to do corporate-to-corporate value transfers. A corporate treasurer could keep a large stash of Mondex value in her workstation and zap transfers directly to counterparties. Mondex could also be used for instantaneous foreign exchange, replacing promise trading and settling up at the end of the day, which exposes a company to the risks of currency fluctuation.

7 DIGICASH

The history of computing is dotted with individuals who left their permanent mark on the development of the Information Age. Some, like John Atanassoff, who constructed the first semielectronic digital computing device in 1939, remain almost anonymous. Others, like Bill Gates, Mitch Kapor, Steve Jobs, and Steve Wozniak have become household names—in some households, anyway. When the history of digital dough is written, Dr. David Chaum, the founder of DigiCash, may be in the pantheon of people who played a crucial role in making cyberbucks a reality.

The Network Without Eyes

Dr. Chaum, who received his Ph.D. in computer science from the University of California at Berkeley, and his colleagues at the Center for Mathematics and Computer Science, a government-funded institution in Amsterdam, have developed something called a blind signature, which allows numbers to serve as electronic cash.

Dr. David Chaum

To understand the key role blind signatures play in the development of digital cash, we need to step back and take a look at just how much information is gathered from us every day by third parties.

Every time we make a phone call or buy something with a credit card, buy a magazine subscription or pay our taxes, that information is going into someone's database. By linking that information, someone can create an extremely detailed dossier on your life: who you talk to, what you read, what your income is, where you eat, what you drive. It is impossible for any of us to control this information. We don't know everyone who's keeping a file on us. We don't know if the information in those files is accurate. We don't know who has access to that information.

These files, by themselves, may not be bad. You might even benefit from them. You wouldn't want your bank to put your money at risk by lending it to deadbeats. And it's certainly convenient to receive instant credit at a shop because a bloodless computer somewhere tells the shopkeeper it's okay to let you run a tab.

The flip side to this is the mischief that can be wrought if this information falls into the hands of a bad actor. Credit card thieves can run up balances on your plastic. Murderers have been known to use government records to find their prey. And the Internal Revenue Service has used household income records maintained by mailing-list companies to target taxpayers for audits.

The reason all this information on you can be tied up in a neat package for prying eyes is that everyone in the United States has a convenient identifier: their Social Security number. Identifying everyone by a number is a social tradeoff, a tradeoff by which social control bests the interests of the individual in empowerment and privacy. What Dr. Chaum's blind signature does is make this tradeoff unnecessary. It allows individuals to perform transactions anonymously while protecting society from fraud.

"In our system, people would in effect give a different (but definitively verifiable) pseudonym to every organization they do business with and so make dossiers impossible," Dr. Chaum wrote in the August 1992 issue of *Scientific American*. "They could pay for goods in an untraceable electronic cash or present digital credentials that serve the function of a banking passbook, driver's license or voter registration card without revealing their identity. At the same time, organizations would benefit from increased security and lower record-keeping costs."

According to Dr. Chaum, the hardware already exists to implement this system. Intelligent agents—software programs used to "run errands" for computer users—can store and manage their owner's pseudonyms, credentials, and cash. And smart cards have the computing power to run the algorithms necessary to make the system work.

A digital signature uses cryptography to assure the integrity of electronic messages. A person creates a message and "signs" it using a private key. The private key is known only to the creator of the message. The message is sent to a recipient who possesses a public key. The public key is given by the creator of the message to his or her correspondents. It is used by the recipient to verify the signature on the message. Trying to forge these electronic signatures is a daunting task. "The best methods known for producing forged signatures would require many years, even using computers billions of times faster than those now available," writes Dr. Chaum.

Now, if we take these digital signatures and apply them to cyberbucks, here's how they might work.

Digital notes, or messages signed using a private key, would be issued by a cyberbank. Different keys could be used for various denominations of cash: one key for $1, another for $5, and so forth. The banknotes could be verified by a public key from the bank. A depositor at the cyberbank could generate a note number (in this system, users, not the U.S. Treasury, would number the dollars). The number would be a random 100-digit number to make negligible the chances of anyone else generating the same number. The depositor signs the number with his or her private key, corresponding to her digital pseudonym. The depositor's public key would be on file with the bank. When the depositor's note arrived at the bank, the bank would verify the depositor's signature, strip it from the note number, sign the note with a signature for the requested denomination, and debit the depositor's account. Then the note with the bank's signature and a withdrawal slip are sent to the depositor.

When the depositor goes to make a retail purchase, she slips her smart card into the merchant's terminal and transfers one of the signed banknotes to it. The terminal verifies the signature, then sends it to the bank

> **Identifying everyone by a number is a social tradeoff, a tradeoff where social control bests the interests of the individual in his or her empowerment and privacy. What Dr. Chaum's blind signature does is make this tradeoff unnecessary.**

> "In our system, people would in effect give a different (but definitively verifiable) pseudonym to every organization they do business with and so make dossiers impossible."
> Dr. Chaum

for reverification. The bank checks the note against its list of issued notes, credits the merchant's account, and sends him a deposit slip. The depositor is given her goods plus an electronic receipt with the merchant's signature, and leaves the shop.

All the transactions in this example are protected. The signatures used by each of the parties prevent anyone from being shortchanged. The merchant can't deny that he received payment; the bank can't claim that it didn't issue notes or that it didn't receive them from the merchant for deposit; the consumer can't say that she didn't withdraw the notes from her account, nor can she spend the notes twice.

As secure as this system is, it still isn't very private. The bank can identify who the depositor was in these transactions by keeping tabs on the numbers of the notes it issues. As soon as the merchant deposits the note, the bank will know where and when the depositor spent the money—something that can be done with credit cards but not with hard cash.

Moreover, files based on digital signatures are more vulnerable to abuse than conventional files because a person can authenticate the information in the files without giving the data away or revealing their source. That person, for example, could determine that the depositor made purchases from the merchant on ten occasions, although he would be unable to determine what the depositor bought and when she bought it. So while the actual value is protected in the face of an assault by a hacker, the hacker can still find data about buying habits.

What Dr. Chaum has developed is a way to ensure privacy in these kinds of transactions. It works like this.

Before sending the note number to the bank, the depositor multiplies the number by a random factor. When the bank gets the depositor's message, it doesn't know what it's signing; it only knows the message carries the depositor's signature. After receiving the blinded note from the bank, the depositor divides out the random multiplier and uses the note as before.

The blinded note is untraceable. Even if the bank and merchant collude, there's no way they can determine the multiplier, so they can't identify who spent what note. But if the depositor loses the note or it's stolen from him, he can reveal the multiplier, so the bank can stop the note or trace it.

The problem with this system is that the note numbers can be easily copied. To prevent double spending, each note must be compared online to a list of notes when it is spent. This may be okay when large sums are involved, but is too expensive for small transactions, such as buying a jelly donut for breakfast.

To address this problem, Dr. Chaum and two of his colleagues, Amos Fiat and Moni Naor, have devised a system by which the payer answers a random query about each note when he spends it. Only the user knows the answer to the query, ensuring that a spoofed e-mail address won't lead to fraud. Answering the query doesn't compromise the payer's privacy, but if someone tries to spend the note again, it can be easily traceable to the payer's account.

Dr. Chaum's company, DigiCash, has created smart cards that use this scheme. The cards are used in two buildings in Amsterdam; workers can use the cards to pay for faxes, food in the cafeteria, and coffee in vending machines.

A Vehicle for Ecash

Dr. Chaum took his blind-signature work and founded Digicash Corp. in the Dutch capital in 1990. He says the company is making money.

Unlike Mondex, DigiCash has more than one iron in the fire. In fact, its first project had nothing to do with digital money. It was a road toll system developed for the Dutch government. It is now being tested in Japan and marketed in other countries.

Amtech, a major player in automatic road toll collection and the sole supplier of microwave tagging for all North American and European rail, is a nonexclusive licensee of DigiCash's high-speed digital cash payment technology for road toll applications. DigiCash is fabricating its own chip cards and various readers and interfacing them with Amtech's microwave technology. DigiCash's concern for privacy extends to these toll systems, too. Its toll system allows a secure payment from an IC card while keeping the roadside equipment from identifying the card.

DigiCash first demonstrated how to make payments over the Internet at a conference in Geneva long before some of its competitors, such as Cybercash, First Virtual Holdings, and Open Market, were a pulse on the Web.

According to DigiCash, many thousands of people are using e-cash in their worldwide trials. Over seventy merchants have joined the trial, from Encyclopaedia Britannica to Ricky's Junk Shop, and accept e-cash. The digital money used in the trial, the Cyberbuck, cannot be changed into real money, but valuable goods and services can be purchased with it.

Among DigiCash's products are Ecash, smart-card technology for electronic purses and wallets, and about eleven patents likely to assure a stream of licensing income as electronic commerce evolves.

DigiCash's Home Page on the Web

Ecash

Ecash is DigiCash's name for its brand of electronic lettuce.

Ecash isn't designed to be a substitute for credit cards. When it's displayed on a computer screen, it appears as coins. With Ecash you can pay for access to a database, buy software or a newsletter by e-mail, play a computer game over the net, receive $5 owed to you by a friend, or just order a pizza.

Ecash would be issued by "banks"—either real or virtual—which could certify that coins used to pay for something have not been illicitly duplicated.

Ecash is designed for secure payments from any personal computer to any other workstation, over e-mail or Internet. Ecash has the privacy of paper cash, while achieving the high security required for electronic network environments. Public key digital signature techniques protect against attempts to intercept the digital money, as the Ecash flows to its destination over the Internet (or any other computer network).

With the Ecash client software a customer withdraws digital money (Ecash) from a bank and stores it on his local computer. The user can spend the digital money at any shop accepting Ecash, without the trouble of having to open an account there first, or having to transmit credit card numbers. Because the received Ecash is the value involved with the transaction, shops can instantly provide the goods or services requested and don't have to jump through hoops to accept payment, such as the merchant accounts needed to accept credit card payments. Anybody who runs the free client software can both pay and be paid by any other Internet user running the software.

Another security feature of Ecash is the protection of Ecash withdrawals from your account with a password that is known only to you, not even to your bank.

Ecash is also the only Internet payment system that protects the payer's privacy with technology in the user's software rather than with the unverifiable policy of a distant company. DigiCash's blind-signature technology allows a bank to securely protect itself against fraud while preventing it from collecting dossiers on a customer's purchasing habits. This privacy protection is not available with systems based on credit cards, ATM/POS, or electronic checks.

DigiCash makes its Ecash software available free to users and merchants, although banks may need some special encryption hardware for speed and extra security. It can be downloaded from the company's Web site.

Currently, DigiCash makes graphic versions of its Ecash software for Microsoft Windows, a number of X-Windows platforms, and the Macintosh. There are also text versions available for these platforms, as well as several other Unix platforms. A TCP/IP network connection is always required.

In addition to working on the World Wide Web, Ecash will also work with FTP or Gopher. And versions of the software released in June 1995 support for embedding Ecash messages in e-mail. This software does the work and then places a coded message in the e-mail.

In general, Ecash works like traveler's checks. When you lose Ecash you have to report the loss to the bank and supply it with the serial numbers of the lost coins. (Because of the built-in privacy in the system the bank does not know which coins were supplied to you.) These serial numbers allow the bank to check if the coins are really lost and to refund the money.

How are Ecash accounts protected? The Ecash executable generates its own private/public key pair; the public keys are announced to the bank when a depositor opens an account. The only way to access a bank account is to use the private keys, which never have to leave a depositor's computer.

Ecash payments can't be traced. The underlying Ecash protocol protects the privacy of the customer so that banks cannot create a list of descriptions or amounts of purchases made by payers. Each coin has its own note number (random, determined by your Ecash executable, not by the bank), which has been signed by the bank, but in such a way that it is impossible for the bank to know which note numbers it has signed.

Shops may be able to track their customers' purchases by user and host name, but this is due to the default communications system used (the World Wide Web, which usually logs connections), and because of the need for shops to send goods and receipts remotely to their customers, not because of the underlying Ecash protocol. If a payment was sent by anonymous remailer, the payee would not be able to trace the source of the payment to either a host name or bank account.

Payments are not private for payees. Ecash banks are able to list the amounts of all payments received for all accounts. A payer knows to what account he has paid, and typically gets a receipt proving payment to that account.

The shop doesn't know who a payer is, but it does know his host ID (otherwise, it wouldn't be able to talk to the payer in the first place). If a payer really wants to prove he is paying, he can decide to give up his anonymity for a particular payment and include his identity (e-mail address) in the description linked to that payment.

While Mondex is still working on its Internet product, DigiCash already has the means to combine the network form of Ecash with smart cards so the money can be taken off the network and stored in the card's chip for offline or point-of-sale purchases.

DigiCash's Ecash Home Page: Electronic Money With All the Advantages of Real Cash

http://www.digicash.com/ecash/ecash-home.html

Electronic Wallets

DigiCash is moving into the electronic-wallet market through the Conditional Access for Europe, or CAFE, project, an initiative of the European Community to develop a secure electronic payment system that protects the privacy of the user. There are some thirteen partners from several countries involved. The target is to make electronic wallets that can be used for payment, access to information services, and—if required—identification. It will be an open but secure system.

In the future, it may be broadened to include electronic personal credentials (like passports, driver's licenses, or house keys) and medical information.

The hardware will exist of pocket-sized electronic wallets. Several versions will exist—some simple, with just two buttons, some with large LCD screens and lots of buttons. Some may be extensions to existing personal digital assistants (PDAs).

The wallets will use smart cards as money storage devices. They will have an infrared interface to make them easy to use. It will be possible to make user-to-user payments.

DigiCash is working on the project with a consortium of firms that

Presenting Digital Cash

includes Gemplus, France Telecom, and PTT Netherlands. A trial card-and-electronic-wallet system is expected to be up and running in the third quarter of 1995 in the headquarters buildings of the European Commission in Brussels.

CAFE–A Pan-European Project

http://www.digicash.com/products/projects/cafe.html

Have a look at the future.

- Introduction
- List of participants
- How to get the software simulation
- For more information

CAFE, the electronic wallet for the information age

- CAFE lets transactions be as private as cash payment are today and transfers electronic money that's more secure than today's best bank notes.
- The CAFE wallet protects the issuer against fraud and the holder against loss.
- CAFE can check a person's authority to have access to restricted information, services and areas, while maintaining his privacy.
- CAFE combines high security with speed and consumer convenience.
- CAFE will open possibilities to banks, travel and telecommunications operators, information services, health care, customs and security firms, and many other services.

A pan-European device

CAFE is an European project, carried out by a consortium of companies active in electronic payments together with leading research organisations. It is supported, also financially, by the European Commission. CAFE is an acronym of Conditional Access For Europe, and its name reflects the scope of the project. It is developing an electronic wallet, to be used as a pan-European device for consumer payments, access to information services and - if required - identification. The project will conclude with a trial of the payment device in the second quarter of 1995. After a successful trial with actual users the technology should be ready for the market in 1996.

Public key cryptography

The CAFE system is based on recent research in public key cryptography. The combination of a public key and a unique private key makes a new generation of smart card and wallets possible, that uses high security protocols while safeguarding the privacy of individuals.

Smart Cards

From the day it was founded, DigiCash has been developing smart-card mask technology.

DigiCash has developed many smart-card masks for its own products and the products of others. The company has been able to bring the cost of these masks down and give smart-card makers more bang for their buck in the security department.

By employing a technology referred by its code name, Blue, DigiCash brought the cost of a chip, with public-key-based stored value, down to under a dollar.

Dr. Chaum, in a statement announcing the release of Blue, observes: "It's generally agreed that off-line transactions like credit card and stored value need public-key cryptography for security; but since public-key chips are inherently more complex and significantly more costly, most systems have been built with less secure cryptography and may need to be redone. Now we have provided a way to use public key on the least expensive and most proven chips available."

Blue obtains its economy through a minimal requirement for silicon. It is currently implemented as firmware for the micro-controller chips produced in greatest volume: Motorola SC-24 and SGS-Thompson ST301/601, with masks for other silicon suppliers under discussion. Blue requires only 1k bytes (the smallest configuration available) of EEPROM memory, the main factor in the cost of chips for smart cards.

Blue has other significant advantages. Most chips, for instance, can irrevocably scramble the valuable data they store when power is interrupted unexpectedly—for example, during a power failure or when a user removes the card too early—but Blue is able to protect the chip's data.

Other cards reveal the card identity and data content to any reader or anyone tapping communications. Blue, however, encrypts everything communicated while revealing only necessary information, and only to readers with corresponding keys.

Another smart-card product from DigiCash is its "Sake" card. This smart card is geared toward giving its customers more flexibility. Where other cards force the use of protocols fixed by the card's vendor and require an application to be based on a specific file-organization model, the Sake card allows the design more flexibility.

The application is programmed into the Sake card itself. Since it doesn't have to operate from the card reader alone, there is a better grip on functionality and security. Customer protocols or security measures can easily be programmed directly into the card.

DigiCash licensees can buy the company's chips directly from multiple suppliers of silicon, such as Motorola and Thompson. The chips can then be made into cards by any one of more than a dozen companies around the world today.

DigiCash has also licensed its smart-card technology to MasterCard for a demonstration system implementing the first smart-card chip mask technology meeting the latest EMV (Europay, MasterCard, Visa) standard. These advances provide true dynamic public key authentication using the least expensive smart-card chip available. DigiCash designed cards that included prepaid cash replacement functions and loyalty systems, access control, and other applications for MasterCard's credit/debit

By employing a technology referred by its code name Blue, DigiCash brought the cost of a chip, with public-key based stored value, down to under a dollar.

systems. DigiCash's advances in high security and reliability are integrated into the design. DigiCash developed a complete system including terminals, PIN pads, host computer, and all related software. Cooperation is ongoing.

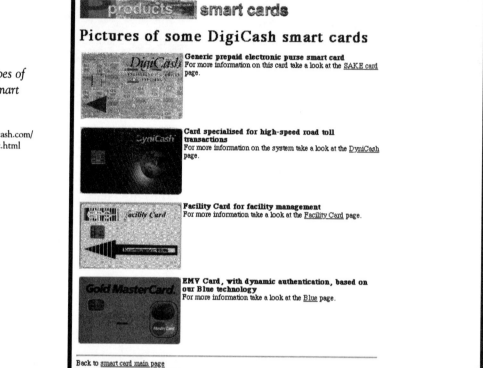

Various Types of DigiCash Smart Cards

http://www.digicash.com/products/cardpic.html

Facility Cards

In addition to making smart cards for general e-cash purposes, DigiCash makes "facility" cards for more narrow digital cash applications.

Facility cards are ideal for a closed system, in which the owner of the cards is also the owner of the devices accepting the cards. For example, a corporation could use the cards for copy machines, soft drink vending, even parking. The cards make it easy to track usage, and can also be used as an effective tool for client billing and resource allocation.

Facility cards are now being used in schools, hospitals, industry, offices, and recreational sites. Some advantages of facility cards:

- Facility managers can obtain information and control, including the separation of access control security from payments and other functions.

- Cash handling is reduced or eliminated.

- The cards can be seamlessly upgraded to accommodate changes in a facility's management.

- The cards work with almost any equipment—vending machines, photocopiers, fax machines—and interface with all access control systems and electronic cash registers.

- Card readers can stand alone or reside on a network.

- The cards are fully reprogrammable and the software can be securely downloaded.

Pre-paid Smart Cards for Specific Uses in a Closed System

http://www.digicash.com/products/projects/fcard

products → facility card

One card provides a complete facility management solution

- What is Facility card?
- Why Facility Card different
- Hardware see the Facility Card hardware page
- Value added resellers
- An example of Facility Card versatility

What is Facility card?

It's an advanced prepaid smart card system for closed site cash replacement; it securely automates vending machine, point of sale, phone and photocopier payments, access control, time/attendance systems and other services. Budgeted individual discounts and access rights are easy, extensive audit information is accumulated automatically. It is now operational in schools, hospitals, industry, offices, and recreational sites.

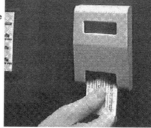

Facility card advantages

- *Management and control flexibility*- Facility managers obtain information and control, including separation of access control security from payments and other functions.
- *Eliminates cash handling*- Whether at attended or unattended points of payment, automating cuts through all the costs and problems of cash handling.
- *Grows with your needs*- Any Facility card can seemlessly be upgraded step-by-step to incorporate all the features of a complete management system.
- *Works with almost any equipment*- Controls almost all modern vending machines, photocopiers, fax machines, and interfaces to all access control systems and electronic cash registers.
- *Offline and/or Online*- Card readers can stand alone, connected only by the occasional visit of our data acquisition terminal, or by various networking options.
- *Fully reprogrammable*- A secure software download capability, standard on each reader, makes for a sound investment in the future of chip cards.
- *Upgradable to open systems*- Drawing on our technological expertise in the emerging national chip card systems, upgrade to compatibility to such systems is provided for.

The Missing Piece

Despite all DigiCash's whiz-bang technology, it has failed to persuade any banks to join its grand Ecash strategy.

Dr. Chaum is now spending a lot of energy trying to convince banks of their shortsightedness. His approach is to try to sell the financial institutions on what he sees as the changing paradigm in the commercial marketplace. That change, according to Chaum, will result in a new model of commerce in which power shifts toward the individual, who can choose with whom to do business from among millions of potential "counterparties."

The great leveler in the Information Age has been the personal computer. More and more people are discovering this. They are finding their power augmented by the PC and are hungry for ways to multiply that power even more. Anyone who can feed that hunger, Dr. Chaum reasons, can make a lot of money. He contends that most corporations today are failing to seize all the opportunities in wired commerce.

Banks are uniquely placed "to take the technological possibilities in hand, create a market, and steal the show," Dr. Chaum told *American Banker*. "I tell bankers that if they offered customers something that looked after their interests, they might be able to create the kind of relationship that Apple got with its Macintosh. Some people just love that company, and that's not something you see in banking or credit cards."

He told the banking publication: "Bankers like to say they have a monopoly on the payment system, that it is the main advantage of their franchise, and that their principal asset is the customer relationship. Digital cash can add value and enhance the relationship, and we are the only ones with digital cash."

Without bank backing, DigiCash is in danger of becoming a bit player in a market where it has been a pioneer. Secure credit card transactions will be the most competitive part of the electronic commerce market, a market where a company's partners will be crucial to its success—and one of those partners had better be a bank. Without one, a company may end up being a back-alley boutique on the infobahn.

> "I tell bankers that if they offered customers something that looked after their interests, they might be able to create the kind of relationship that Apple got with its Macintosh."
> Dr. Chaum

8 CYBERCASH

CyberCash has taken a very different route than DigiCash. Instead of investing a great deal of time in technology and user interface, they've worked with the existing infrastructure and the major players in the industry. The importance of banks to the success of any company that wants a respectable chunk of the cyberbuck market won't be lost on CyberCash. While Dr. Chaum talks the talk of the technology visionary, the crew at CyberCash know how to talk the talk of bankers.

The CyberCash management team is deep in talent and large in scale. At the point of CyberCash's leadership pyramid is CEO William N. Melton. CyberCash is his fourth entrepreneurial venture in this arena.

Melton knows a little bit about the concerns of banks when it comes to performing transactions in the electronic ether. He founded VeriFone, the company that makes the devices retailers use to authorize credit card charges. Verifone was able to manufacture and sell compact terminals for $500 each. It has since grown to $300 million in annual sales, with 4 million devices installed worldwide.

What VeriFone terminals do—capture and transmit card and transaction data—is much like what CyberCash's software promises to deliver for transaction data entered into personal computers. The similarities do not end there. Neither VeriFone nor CyberCash is out to make great advances in technology; both focus on applying available hardware or software to handling payments.

Before founding VeriFone in 1983, Melton, in 1971, founded Real-Share, a database and telecommunications company that served the majority of Telecheck franchisees. Through Real-Share, he pioneered the use of minicomputers, voice response systems, and distributed processing nodes in the financial community.

In 1991 Melton helped launch Transaction Network Services (TNS), a data transmission network for six of the twelve largest credit card processing centers. TNS, which made a highly successful initial public offering in the spring of 1994, currently handles more than 3 million financial transactions daily. Melton still serves on the board of TNS, and the company is expected to be part of CyberCash's private banking network.

In addition to his continued service on the boards of both VeriFone and TNS, Melton also is a member of the board of directors of America Online.

Given his experience in setting up infrastructures, Melton is keenly aware that the reputations and relationships of the top people in the organization are critical to acceptance by the industry at large. Sometimes who you know is as important as what you know. Technology and ease of use aren't going to determine who wins the digital cash wars.

It's worth spending a minute to look at the extraordinary depth of CyberCash's management—it will go a long way toward affecting the company's prospects.

The chairman of the board of directors is Daniel C. Lynch, who is also one of the company's founders. Lynch was chairman and founder of the Interop Company, which is now a division of Softbank Expos and was formerly the Ziff-Davis Conference and Exhibition Company, in Foster City, California.

As the director of information processing for the Information Sciences Institute in Marina del Rey, Lynch led the ARPANET team that made the transition from the original NCP protocols to the current TCP/IP-based protocols. Lynch directed this effort from 1980 until 1983.

The day-to-day management of CyberCash is done by chief operating officer Bruce G. Wilson. Wilson, another founder of the company, has more than twenty years' experience in the software and telecommunications industries in engineering, strategic planning, marketing, sales, and executive management positions. Previously he was vice president for electronic funds transfer systems at NYNEX and was instrumental in launching its interactive videotex unit and software systems integration business units.

Wilson was elected to the board of the Electronic Funds Transfer Association (EFTA) and is chairman of its finance committee. He founded and was the first chairman of the EFTA Prepaid Card (smart card) special interest group.

Another old pro in the upper echelons of CyberCash is Dr. Stephen D. Crocker, who is also a founder of the firm and its senior vice president for development, responsible for security architecture and the design and implementation of the CyberCashcher at USC Information Services Institute.

Dr. Crocker was part of the team that developed the original protocols for the ARPANET. He has served as the area director for security in the Internet engineering task force for four years and is now a member of the Internet architecture board.

CyberCash's Approach

CyberCash was founded as a privately held company in Vienna, Virginia in 1994 with the aim of getting banks to participate in Internet commerce. It compares what it's trying to do with what companies like National Bancard offer to banks and conventional retailers today. The difference is that all the merchants on the CyberCash program would sell their wares by interactive computer networks. Before CyberCash can sell banks on the idea, however, it will have to ensure it can deliver secure transactions on the mother of all nets.

CyberCash's Home Page

http://www.cybercash.com

"There is no security on the Internet," Dan Schutzer, president of the Financial Services Technology Consortium, a group of major banks, told the *Wall Street Journal*. "Your conversations can be tapped, your passwords can be obtained, and your credit card number can be filched. Clearly, it's there for the reading for a clever hacker."

As we discussed before, the Internet now is primarily for window shoppers. There are hundreds of products out there—T-shirts, books, compact disks, rope sandals, legal services—and hundreds of companies have erected electronic storefronts—travel agencies, art galleries, real estate brokers, a Volvo dealership—but, for the most part, all a void warrior can do is read product descriptions. If he wants to buy something, he has to call the vendor by phone. This cumbersome fact of life has hurt Internet commerce, estimated by *Internet Letter*, a newsletter aimed at business users, to be worth the paltry sum of $10 million a year.

A few pioneers have passed credit card numbers safely on the Internet. An electronic bookstore has received payment over the network. A small start-up retailer in Nashua, New Hampshire, sells compact disks online. A few banks offer services in which consumers can type their credit card numbers into a computer, but there is a risk that the number can be captured by an outside party or that the retailer is unethical and will use the number illegally. So these banks are exceptions to the rule. Almost no one has been able to bring banks directly online to make automatic payments to Net merchants, which is CyberCash's goal.

Under the company's scheme, a consumer can establish a CyberCash account—a non-interest-bearing holding account for cash—or a credit or debit card. Consumers can go online, click a "buy" icon, and type in their account or credit/debit card number. Frequent users may preregister numbers to avoid typing in the data with every purchase. The number would be encrypted at the computer and a code would be sent to CyberCash, which would unscramble the message and present the card number to the issuing bank for authorization. Only CyberCash and the issuing bank would have access to the codes to read the card numbers. The merchant would receive notice that the transaction had been approved but would never see the credit card number. CyberCash would arrange for settlement through the issuing and acquiring banks.

Even if this scheme calms the qualms of banks, CyberCash may find additional resistance at the merchant end of things because transactions through the system are still riskier than a run-of-the-mill point-of-purchase buy. When someone buys something at a shop and plunks down his credit card, the merchant can see the card. There's a measure of security in that sight-check that merchants may be reluctant to relinquish. However, mail-order houses will probably take this system in stride, since they never get to look at a credit card when someone buys something from them, anyway.

Almost no one has been able to bring banks directly online to make automatic payments to Netmerchants, which is CyberCash's goal.

For its part in the transaction, CyberCash will charge the banks a small fee. The per-transaction fee is expected to be anywhere from ten to fifty cents and could be slightly higher than what acquirers pay today to electronically process a conventional credit or debit card transaction. CyberCash says that the higher fee may result from the additional technology and security required by the transaction.

CyberCash plans to spend $20 million on a private network of computers, which will separate Internet merchants from users' bank accounts.

The company will sell its processing service directly to banks acquiring debit and credit cards. Those banks sign up retailers that want to accept credit card and debit card payment for goods sold over computer networks. The encryption that allows consumers to make purchases would be written into the basic software consumers use to access computer network services, at no extra cost to the consumers.

To protect sensitive account information, CyberCash will use the Digital Encryption Standard (DES), used in most electronic payment applications today, in conjunction with public key encryption, developed by RSA Data Securities. Only persons who have a special software "key" can read the encrypted information. Customers and their banks will hold the keys.

One drawback to teaming up with RSA is that some of its encryption products have been under severe export restrictions in the United States. Those products, because of their perceived threat to national security, have been assigned to the munitions list by the Department of State and the National Security Agency (NSA). But CyberCash has vaulted that hurdle. The Department of Commerce has granted the company permission to export its Internet software for encrypting credit card information even though the software uses the powerful DES and a 768-bit RSA Data Security algorithm.

CyberCash's breakthrough was hailed by Internet watchers as a precedent that will spur electronic commerce. They've interpreted the ruling to mean that the government's top code makers and breakers at the NSA are saying they want to have control over applications that pose a threat, but also want to make electronic commerce easier for business. Free from worries about violating export controls, CyberCash immediately posted its free CyberCash client software, which works with any World Wide Web browser to encrypt credit card information, up on its Web server at http:// www.cybercash.com.

A List of Merchants Who Currently Accept CyberCash

http://www.cybercash.com/merchants/merchant_list.html

Check out these interesting merchants now! They accept CyberCash protected credit card payments today!

Our Newest Merchants:

HotWired

    Industry news, art, conversation and commerce.

Global Network Partners

    Helping people establish successful home-based businesses.

Virtual Vineyards

    Fine California wines and our first merchant!

LifeLink

    Disaster and emergency preparedness supplies.

Silver Cloud Sports, Inc.

    Distributor of high quality, yet affordable Silver Cloud golf clubs.

CD-R Products

    Products and training dedicated to the CD recordable market.

Newer Technology

    Quality memory, accelerator, networking and portable peripheral products.

A Walk Through a Digital Haberdasher

CyberCash has a prototype electronic storefront called Webthreads on the World Wide Web (http://pay.fef.com/~web/).

When you access the Webthreads home page, CyberCash is briefly explained to you. By clicking on "Come shopping in our simple (pretend) store," you can enter Webthreads.

On the "begin shopping" page, you can see the Webthreads unconditional guarantee: "We guarantee everything we sell. If you are not satisfied with one of [the] products at the time you receive it, or if one of our products does not perform to your satisfaction, please contact us either by phone or e-mail for a repair, replacement, or refund." And you can click on "Special Sales & Offers" to travel to the next Web page.

The specials for this month are sweaters. To read the descriptions of these products and see a photo of them, you click on one of four options: chamonix guide sweater, lambswool zip cardigan, wool cable sweater, and ribbed lambswool cardigan.

When you access the product description, you can choose to buy the product or to continue to window-shop. If you buy an item, it's placed in your shopping basket. As you shop, you can check your shopping basket at any time. When you've finished shopping, you pay for your goodies with a credit card and they'll be sent to you.

Webthreads is a very simple prototype, but it gets the point across to prospective buyers.

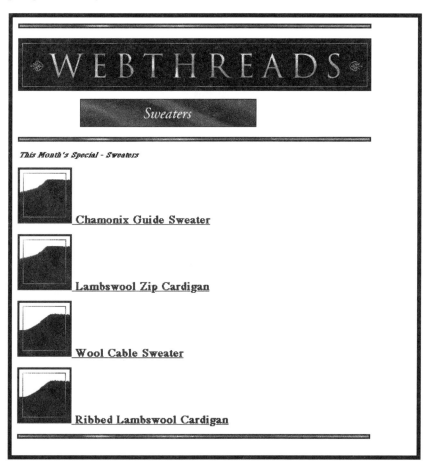

Sweaters Offered by WebThreads, CyberCash's Fictional Merchant

http://pay.fef.com/~web/wtbin/shopping.cgi/sweaters

Digital Cash

CyberCash also is seeking to go beyond credit and debit card transactions and later this year plans to offer true digital cash.

In the CyberCash system, participating banks would let a customer

> **DigiCash's tokens operate like Traveler's checks, while CyberCash's digital money will be based on cash in the bank.**

open cybercash accounts, or "electronic purses." Using the company's software, a customer would move money from the checking account into the electronic purse. As with an ATM, the customer could then withdraw digital tokens from the purse and use them to make purchases on the Internet. Upon receipt, the seller would query the CyberCash computer to verify the token was valid and instruct CyberCash where to deposit the money.

CyberCash's money approach is a little different from that of DigiCash. DigiCash's tokens operate like Traveler's checks, while CyberCash's digital money will be based on cash in the bank. CyberCash will function as an escrow account. When money is transferred, CyberCash sends the information to the necessary banks. Its system is designed to work with the existing banking system in a way that DigiCash's system is not.

Let's Make a Deal

As pressure has increased to get commerce on the Internet rolling, CyberCash has been able to cut some deals that should help it extend its reach through the Net. In the spring of 1995 CyberCash announced partnerships with seven networking software vendors.

The company says the deals with FTP Software, Frontier Technologies, InterCon Systems, NETCOM On-Line Communication Services, Network Computing Devices, Open Market, and Quarterdeck will enable its secure online payment system to be independent of the browser's and user's interface. Because they are working with a broad spectrum of suppliers, this system can become ubiquitous.

CyberCash's Melton said that "each of CyberCash's partners offers a high level of value for the Internet market today and will work with us to provide the added functionality of electronic payment services that users will rely on. Electronic payment users demand a high level of interoperability, security and convenience. CyberCash is committed to meeting these needs through its extraordinary technology, service and industry partnerships."

Each of the seven technology partners has agreed to integrate its technology with CyberCash so that the company's electronic payment services can be easily accessed through each partner's products. A user with a browser supplied by any of these vendors will now be able to make online payment transactions via CyberCash directly from her browser without having to download CyberCash software separately.

In addition to making important inroads in the software arena, CyberCash has enlisted an important hardware maker, Sun Microsystems, to its cause. As part of its relationship with CyberCash, the workstation maker has agreed to begin promoting CyberCash electronic payment services to its customer base. Sun will comarket CyberCash Internet Payment Services and begin educating its customers about the benefits of secure, end-to-end payment services. Sun's servers are utilized by companies that seek to sell goods and services on the Internet.

Almost any vendor with a home page on the World Wide Web could obtain merchant status and begin accepting payment online via CyberCash's payment system. Since CyberCash's software is independent of browser, server, and platform, merchants that rely on Sun's servers can easily conduct transactions with consumers no matter what type of client or browser they may have.

And even competing vendors have found CyberCash an attractive partner. CheckFree, a major provider of electronic commerce services, announced in the summer of 1995 that it was integrating CyberCash's security features and ability to conduct cash transactions into the CheckFree Wallet, creating a single solution for electronic payment transactions that offers checks, credit cards, cash and coin, and micropayments.

"Partnering with CyberCash was the logical choice for CheckFree," said Pete Kight, founder and CEO of CheckFree. "CheckFree is committed to leading the way for electronic commerce, and providing plug-and-play Internet transaction solutions. The CheckFree Wallet is now an even more attractive transaction platform for both consumers and merchants."

The CheckFree Wallet was introduced in April 1995 to allow consumers to purchase goods and services from online merchants in a safe, convenient and familiar manner. It does not require prior registration with merchants, and online shoppers pay no fees or transaction charges. With the addition of CyberCash's technology, consumers will be able to transmit credit card or cash payments securely over the Internet, and merchants will receive authorization in real time. In addition, merchants will be able to accept payments from any online consumers, regardless of the server or browser they are using.

The Trust Thing

One key element to CyberCash's success will be banks' confidence that the firm's encryption technology is safe and effective and the staff can be trusted. CyberCash is hoping Melton's reputation will ensure that trust. Some banks have already endorsed CyberCash's approach to Internet payments.

Wells Fargo & Company was the first bank to announce support for CyberCash transactions. Wells already represented two credit card merchants over Internet when it announced its support for CyberCash, but it wanted to expand to other merchants. Wells's two merchants sell wine, books, and gifts over Internet, but in those programs, consumers transmit their card numbers unencrypted to the merchant.

At the beginning of 1995, Wells, which with several other large and innovative banks is a member of CommerceNet, a consortium of companies in many industries exploring ways to use the Internet, joined in a online payment pilot program with CyberCash in the San Francisco area. The test involves about twenty of Well's 26,000 credit card merchants.

Another bank, First National Bank of Omaha, has also joined the CyberCash ranks. It has announced it will set up a CyberCash transaction server to process credit card transactions. The bank says it will encourage merchants currently using the bank's existing credit card network, First of Omaha Merchant Processing, to set up CyberCash merchant servers.

All this concern by CyberCash over wooing banks isn't misplaced. The S&L disaster notwithstanding, people still trust banks with their money. And anything that wants to call itself money is going to be that much more acceptable if there's a bank behind it.

Melton has been quoted as saying: "We've positioned ourselves to work with the banking industry and make sure that if there are heroes in this, it is the banks."

But he's also said: "The Internet is going to happen, with or without the bankers. But the bankers, the bright ones, are going to make this an opportunity."

9 FIRST VIRTUAL

First Virtual Holding's approach to electronic commerce is, in a nutshell, this: Keep it simple.

From First Virtual's point of view, right now there's a lot of noise surrounding buying and selling on the Internet. Many different companies are talking about setting up mechanisms for Internet commerce. Most of these employ complicated software, require complex encryption methods, and connect to a multitude of banks, servers, and currency exchanges.

First Virtual sums up its philosophy in one of the many FAQs—frequently asked question files—it has posted on the Internet:

"Our philosophy is simple: we believe that secure Internet commerce should be broadly available, easy to participate in, and compatible with the ways of using the Internet that people already know, like ordinary e-mail....

"We saw lots of schemes for Internet commerce being developed, but all of them seemed too complicated, or too limited in their scope, or too difficult to learn or use. Or, they seemed to have too little respect for the Internet culture and community, or too little respect for the intelligence and maturity of the average Internet user.

"Anyone who wanted to, we felt, should be able to buy or sell information simply and securely over the Internet, without having to:

- Buy new hardware, or install new software;

- Learn complicated new commands, take a class, or read a long manual;

- Worry about encryption, digital signatures, or other complex security measures;

- Meet complex financial requirements, like putting up a deposit, proving a long credit history, or signing a restrictive contract.

First Virtual Holdings

http://www.fv.com/

"In particular, we felt that sellers are the best judge of what their products are worth, and should be able to set their own prices. We also felt that buyers are the best judge of what kind of information they're looking for, and should have the chance to examine information before deciding whether to keep it and pay for it."

The company reasons that if Internet commerce isn't simple enough for ordinary people to use, most people won't bother with it. And ordinary people seem to be an important concern for First Virtual.

As the "Information Superhighway" begins to take shape, the company explains, large companies like Time Warner are making deals with commercial online services like Prodigy, America Online, and CompuServe to sell information, products, and services to subscribers. And they're already reaping a windfall from this new kind of business.

First Virtual contends that individuals and small businesses are being left out of the picture. Allying with one of the existing commercial online services is prohibitively expensive for small businesses, especially those with a narrowly targeted market. And an individual or home business with a small line of information products to sell can't even think of approaching the giant online service providers, who depend on broad mass appeal to keep their subscribers connected.

Almost everything available on these commercial online services is also available, or is becoming available, on the Internet, which reaches many millions more people than all the commercial services put together. The Internet represents an immense community of people around the world who have grown to respect individual creativity and initiative, and would seem to be the perfect place for small, innovative merchants to find an audience and a market for interesting information and media products.

But that's only beginning to happen now, and very slowly, says First Virtual. That's because something has been missing from the Internet equation. What's been missing, according to the company, is a way for buyers and sellers to come together over the Internet and exchange information for payment, without disturbing the unique Internet culture—a culture that places free access to information at the pinnacle of community values.

One of First Virtual's most controversial stances is on the role of encryption in electronic commerce. Simply put, it doesn't believe in it. It contends that encryption is still too complex, difficult to use, and unreliable to be the right answer for everyone.

Encryption is almost always cumbersome and difficult, the company contends. And it always adds an additional step, and something else to worry about. After all, even banks and armored cars are subject to robbery attempts, and sometimes those attempts succeed.

Encryption isn't necessary with First Virtual, the company argues, for a simple reason: When you use its system, sensitive information like a credit card number never has to travel over the Internet at all.

*First Virtual
Clickable Map*

http://www.fv.com/help/
overall_map.html

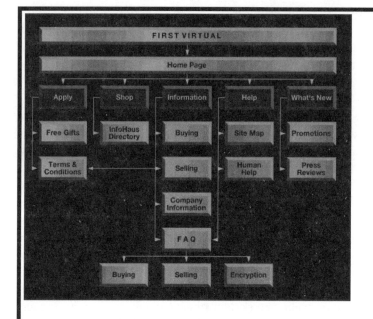

Transactions are all handled with a unique First Virtual account identifier, which may safely travel in ordinary Internet e-mail. Even if the identifier were intercepted, an unauthorized user couldn't use it for fraud. He couldn't even use it to buy information over the Internet fraudulently, because all transactions are confirmed via e-mail before a customer is charged anything. If someone tries to use a First Virtual account identifier to buy something without authorization, the customer can alert the company when he receives the e-mail confirmation message, and the stolen account identifier can be deactivated immediately.

Even when a customer first signs up for a First Virtual account, sensitive information is given to the company over a private phone line. The information is then stored on a computer which isn't on the Internet at all.

But First Virtual's security scheme may not be as bulletproof as it would like its customers to believe. Dr. Frank Rubin of Wappingers Falls, New York, writing to the editors of *DM News* in August 1995, argued that a system like First Virtual's might create greater, not fewer, opportunities for fraud.

"Since I am a cryptographer, and an editor of *Cryptologia*, as well as the owner of a direct mail business, I feel uniquely qualified to comment," Rubin wrote. "The alarming aspect is that creating a virtual bank does not reduce the risk of fraud in transactions over the Internet, it increases the risk.

"First, in addition to the customer's credit card number, there is now a PIN that can be stolen.

"Second, there are now many more points where electronic theft can occur: at the customer's home, between the customer and the local phone exchange, at the phone exchange, between the customer's local exchange and the exchange serving the virtual bank, at the bank's exchange, between the exchange and the virtual bank, between the bank's internal switching unit and the service representative, between the representative's terminal and the computer, at the computer, and outgoing to the vendors.

"Let me illustrate with some plausible scenarios.

"An employee of the virtual bank may print out or transfer to a diskette thousands of records containing names and credit card numbers. An employee may substitute his or her own name and address in records, receive merchandise, then replace the valid records to avoid detection. A thief may tap the phone of just one operator and get hundreds of credit card numbers per day.

"Since the virtual bank holds far more credit card numbers than any one vendor doing business via an 800 number, it would naturally become the focus for theft and fraud attempts.

"I know of no way to provide true security without encryption. The virtual bank simply shifts transactions from the Internet to the phone system, which does nothing for security. Of course, the virtual bank does not actually have to provide security, it only has to convince its clients that it is doing so."

Encryption isn't necessary with First Virtual, the company argues, for a simple reason: when you use its system, sensitive information like a credit card number never has to travel over the Internet at all.

First Virtual Home Page

http://www.fv.com/html/fv_main.html

- APPLY
- SHOP
- INFORMATION
- HELP
- WHAT'S NEW

Three things underlie First Virtual's payment mechanism.

For sellers, the costs associated with selling information over the Internet are much lower than the costs of doing business the traditional way. Duplication costs for information products are virtually zero. The cost of warehousing is trivial. The cost of distribution—once the seller has paid the ordinary cost of an Internet connection—is effectively zero. And returned goods? It costs more to receive returned goods than if the unsatisfied customer just throws them away.

The seller has lost little or nothing by letting a potential buyer examine the information before deciding whether to buy it, even if the buyer decides not to buy. Virtually all of the expense to the seller is due to the cost of developing the information in the first place, and that cost is the same whether you sell ten copies, a thousand copies, or no copies at all.

For buyers, it's often difficult to know whether an information product is worth purchasing without examining or using it. For instance, if someone is selling instructions on installing a computer program properly, a buyer can't know whether those instructions are worth paying for without having a chance to try them out and see if they work. This makes

First Virtual

people reluctant to pay in advance over the Internet for information whose usefulness is unknown.

For everybody, buying and selling have to be simple. Even the most basic encryption methods make Internet commerce so complicated that most people won't bother. And it would be a tragedy for the Internet—which has been built out of the small individual contributions of millions of people—if a few big companies, with the patience and resources to make encryption work for them, came to dominate the Internet and crowd out those individual contributions that have made it such a fascinating place.

In response to these concerns, First Virtual developed a system it feels is both so simple that everyone can understand and use it, and robust enough to be used for almost any kind of information purchase. It's a system, the company says, that lets most buyers try before they buy, and lets sellers set their own prices according to the return they require on their development costs.

By avoiding the need to send sensitive information over the Internet, First Virtual avoids the problems of encryption. Since there's no confidential information being sent, there's nothing that needs to be kept secret.

Under the First Virtual system, if a buyer comes across something that looks interesting, he simply asks for a copy of the information. He provides his First Virtual account identifier to the seller. The seller then forwards information about the transaction, including the buyer's First Virtual account identifier, to First Virtual's Internet payment system server.

First Virtual's server then sends electronic mail to the buyer asking if the information was satisfactory. If the buyer decides that the information is of value, he replies yes, and payment for the item is automatically transferred from the buyer to the seller. If the buyer decides the information is not worth keeping, she replies "no," and is not obligated to pay. Buyers are always asked to confirm their purchases in e-mail before they are charged.

Billing and collections, even when automated, rely on common, easy-to-use Internet tools. A small seller can submit purchase information to First Virtual using e-mail or an ordinary Telnet connection. A large company can easily automate billing and collections.

It costs $2 for a buyer to register with First Virtual. Buyers are never charged a usage fee of any kind. The price set by the seller of a product is the price the buyer pays.

> **For everybody, buying and selling has to be simple. Even the most basic encryption methods make Internet commerce so complicated that most people won't bother.**

Sellers, on the other hand, pay $10 to register with First Virtual, then 29 cents, plus 2 percent of the sales amount on each transaction. Each time a payment is made to their account, sellers pay a $1 processing fee.

From time to time, buyers' credit card accounts are billed for the charges that have accrued during the billing period, and sellers' checking accounts are credited with payment for items sold. First Virtual handles accounting for both buyers and sellers.

The system protects buyers, by giving them a chance to review products and ensure they meet their needs before having to confirm their purchases. At the same time, the system protects sellers by requiring that buyers use this privilege in good faith, and terminating any account-holders who abuse a seller's goodwill. To ensure that customers do not abuse the privilege of trying before they buy, First Virtual may limit the number of times a consumer may evaluate information products without paying for them. This balances each buyer's desire to evaluate before buying with each vendor's need to protect property and predict revenues.

Electronic Data Systems provides support to First Virtual including data processing, documentation, systems engineering, and the electronic settlement of transactions for buyers and sellers. EDS, which is located in Plano, Texas, has also created a way for the merchant to examine a customer's previous purchase activity to make sure there is no pattern of reversals. EDS, which is owned by General Motors, plays a critical role in providing security for First Virtual's transactions. It uses its own network, which is offline from the Internet.

First USA Merchant Services will provide clearing, settlement, and authorization for First Virtual's credit card transactions. Aside from the fact that First USA is affiliated with a credit card company, the First Virtual payment system was created independently of the banking system.

As simple as First Virtual's payment system is, there is a major drawback. While the system lets virtually any person or company set up a storefront on the Internet easily, what can be sold through the storefront is limited to what can be distributed electronically. If someone wants to sell sofas, First Virtual may not be for him. The First Virtual system could be used by individuals selling things that can be digitized—novels, essays, research, a CD-ROM title, photographs, software, or magazine articles.

There's another drawback, one that harks back to why banks may have an upper hand in all this: trust.

Much of First Virtual's scheme turns on trust: A seller ships her goods

First Virtual

with the expectation, but not the guarantee, that she will be paid. For much of what will be initially sold via First Virtual, that uncertainty may not be fatal. A lot of the material would have been given away anyway, so if the author can make a few bucks on something she would have made zilch from without First Virtual, so much the better. This may be a good way to pump some interest into electronic commerce, but it's still miles from a system where transactions are consummated immediately and both parties can feel secure about their dealings.

First Virtual also operates a public-access information server called the InfoHaus, which enables virtually anyone to become an information merchant on the Internet quickly for almost no cost. Using ordinary Internet tools, any seller—from an individual to a huge corporation—can build a customized Internet storefront on InfoHaus, which handles billing, collections, and payment for its sellers automatically.

Any accountholder may become what First Virtual calls an InfoMerchant simply by uploading his information product to the InfoHaus, providing a description of it for consumers, and specifying the sale price.

First Virtual takes care of everything for the merchant: distribution of his information products, confirmation of purchases, accounting, billing, collections, and payment of the proceeds. For those services, InfoHaus takes an additional 8 percent of the transaction price for an item.

Infomerchants can post information products on the InfoHaus, and consumers can browse the products from anywhere on the global Internet, 24 hours a day, 365 days a year. It's like a virtual mall that's always open.

First Virtual's corporate structure is unique, like the product it sells. It has fully embraced the principles of what authors William Davidow and Michael Malone call the virtual corporation. The company owns no real estate or physical offices. It's incorporated in Wyoming; its toll-free phone lines are answered in Oregon; and its computers are in Ohio. Even the company's business cards have e-mail addresses, rather than street addresses. "No two people (in the company) have the same zip code and, for a while, no two people had the same area code," First Virtual president and chief executive officer Lee H. Stein told the San Diego *Business Journal*.

Although in many respects First Virtual is the very model of an Information Age company, it may be taking its virtuality a little too far when it comes to customer service. In one of its FAQs posted on the Internet it advises those of us of the flesh-and-blood persuasion:

> ...what can be sold through the storefront is limited to what can be distributed electronically. If someone wants to sell sofas, First Virtual may not be for him.

Any accountholder may become what First Virtual calls an InfoMerchant simply by uploading his information product to the InfoHaus, providing a description of it for consumers, and specifying the sale price.

"We are trying to provide the Internet with a means of buying and selling that is open to all, and that imposes the minimum possible costs on buyers and sellers. In order to help us keep the cost of the system low, we ask you kindly to please search the FAQ documents, and our World Wide Web server (if you have access to the Web), for the answer to your question before you attempt to reach a human being. Human operators are expensive, and ultimately the consumer inevitably pays for such expenses.

"In the event that the answer to your question isn't contained in that FAQ or the other twelve or so FAQs, or on our Web server, you can reach a human being by sending mail to 'humanhelp@fv.com.'"

Rock Stars and High-tech Gurus

The management team at First Virtual is as distinctive as the company's organization.

CEO Stein, who says he likes to "play on the leading edge," has a past that hasn't been confined to a silicon box or the electronic ether. Stein, the son of a children's summer camp director, was born and raised in Philadelphia. While attending law school at Villanova University, he saw a Merv Griffin show on which a Hollywood producer chatted about his business manager. Business manager sounded like a "cool" occupation to Stein, so, soon after graduating from law school in 1978, he packed his bags and headed to Tinsel Town.

Before long, Stein created his own company and signed his first client, Bo Goldman, the Academy Award–winning screenwriter of *One Flew Over the Cuckoo's Nest*, who later won a Golden Globe for *Scent of a Woman*. Another client was a then-little-known rock band called Men at Work. Within six months of signing with Stein, the group had a hit album that sold millions of copies. In 1980, Stein began working as a business manager to rock stars like Rod Stewart and movie stars like Gene Hackman.

By the mid-1980s, Stein sold his entertainment business and moved to San Diego. He invested in real estate during the boom years, but sold his holdings just before the crash in real estate prices in the '90s.

Stein got involved in First Virtual by happenstance. At an airport, Einar Stefferud, an expert in global messaging systems, was reading his e-mail when Stein, educated enough about technology but not quite sure what Stefferud was doing, introduced himself. The two struck up a conversation and found they were simpatico. They happened to be on the same flight and rearranged their seats so they could sit next to each other.

After a few more e-mail messages and telephone conversations, Stefferud and another Internet expert, Nathaniel S. Borenstein, were off to New York to have dinner at the Hard Rock Café with Stein, singer Peter Gabriel, and investment bankers from Salomon Brothers.

Stein said in a published interview that First Virtual's virtuality made it easy to put together the company's technical team. "We were able to assemble a team of Internet gurus without asking anyone to move," he said. "Since there are no relocation expenses, the decision to become involved was effortless."

The triumvirate of technological experts at First Virtual consists of chief scientist Borenstein, principal Marshall T. Rose, and chief visionary Stefferud.

Dr. Borenstein, who received his Ph.D. in computer science from Carnegie-Mellon University in 1985, is the primary author of MIME, the Internet standard for multimedia and multilingual mail messages. MIME allows Internet users to send binary files—files that contain objects like pictures and sounds—across the Net. His other software credits include metamail, a public-domain MIME implementation in use by millions of Internet users; the Andrew Message System, a pioneering multimedia mail and bulletin board system; and ATOMIC-MAIL and Safe-Tcl, two languages designed to support mail-enabled applications and active messaging.

Dr. Borenstein has also written two books, *Multimedia Applications Development with the Andrew Toolkit* (Prentice-Hall) and *Programming As If People Mattered: Friendly Programs, Software Engineering and Other Noble Delusions* (Princeton University Press). He is a member of the Electronic Frontier Foundation and Computer Professionals for Social Responsibility. He has also served as an advisor to the White House and several international organizations, including the United Nations, the World Bank, and the Institute for Global Communications.

Chosen one of the communications industry's top ten visionaries in 1992 by *Communications Week* magazine, Marshall T. Rose has designed, specified, and implemented protocols and systems in nearly all aspects of Internet technology. He is presently the Internet Engineering Task Force (IETF) area director for network management, one of a dozen people responsible for overseeing the global Internet standardization process.

Dr. Rose, who was awarded a Ph.D. in information and computer science from the University of California–Irvine in 1984, is the author of

At an airport, Einar Stefferud, an expert in global messaging systems, was reading his e-mail when Stein, educated enough about technology but not quite sure what Stefferud was doing, introduced himself.

six books on Internet technology, including network management, electronic mail, and directory services, all published in Prentice-Hall's "Series in Innovative Technology."

Another *Communications Week* Top 10 visionary with First Virtual is the company's resident visionary, Einar A. Stefferud. Stefferud, who received an MBA from UCLA in 1961, can trace his Internet roots back to ARPANET, where he established and moderated one of the first mailing lists on the Net. Over the last twenty years, he has been deeply involved in standards development for the Internet as a member of the Internet Engineering Task Force and other professional groups. In addition to his role at First Virtual, Stefferud is an adjunct professor of information and computer science at the University of California–Irvine.

Spreading the Word and Attracting Allies

Stein and his associates launched First Virtual in October 1994. Within a few hours they had their first customer: a merchant from Helsinki. He was followed by a Dallas software company and a dozen more clients from around the globe.

According to First Virtual, it has signed more than 150 retailers, receiving more than 200,000 hits a day and enjoying sales volume growth of 15 percent a week. Vendors include Apple Computer, Reuters, National Public Radio, the Internet Society, and the music seller Sound Wire. KGNU, a public radio station in Boulder, Colorado, even collected pledges through the firm.

Bigger things are in the wings, according to the company. It expects to branch into catalog shopping with some large name players.

Initially, First Virtual accepted only bank credit cards, but since that time it has managed to net American Express. American Express, which has also announced deals with CyberCash, Netscape, and Open Market, said that as a result of these arrangements, its cardholders will be able to use American Express cards to make purchases, and service establishments will be able to accept payment for goods and services sold over the Internet.

First Virtual's strategy to garner a substantial portion of the credit card business over the Internet could be a very lucrative one. Industry experts estimate that $250 million worth of goods and services was purchased via charge and credit cards over the Internet in 1994.

According to MasterCard data, 66 percent of World Wide Web users use the Internet to shop and 59 percent consider it an important source of product information. Only 28 percent have actually purchased goods or services there—a statistic that First Virtual intends to change.

First Virtual's relationship with EDS has also borne fruit. EDS is selling First Virtual to banks as a way for financial institutions to establish a presence on the Internet so they can provide secure access to their customers for home banking and bill payment services. EDS already provides home banking services to financial institutions through Interactive Transaction Partners, a joint venture with US West and France Telecom. Using the Interactive Transaction Partners platform, banks can offer home banking and bill payment services to customers through touch-tone phones, personal computers, or screen phones.

Through two new services, WebIt and NetIt, banks processing electronic transactions through the joint venture will also be able to offer customers access to the services through the Internet. The World Wide Web will serve as the gateway to the services.

While First Virtual's system is based on credit card transactions, it's bound to influence the development of e-cash systems on the Net, even if it's only to fuel commerce on the Net, make buying and selling more prevalent, and build users' confidence that cyberspace can be a secure place to conduct business.

10 OPEN MARKET

When you log on to Open Market's home page on the Web, one option available to you is a camera view of Boston. Click this option and you can view a photo of the Boston skyline from Open Market's headquarters in Cambridge. A computer updates the photo every minute (the company puts together an MPEG movie of the snapshots every hour and at the end of each day). Why be content with the bloodless lines and numbers on the Weather Channel when you can see for yourself what the weather is like? It's a neat idea. And one that says something about the corporate culture at Open Market, a culture that harkens back to the early days of the PC business.

To discourage long, yawn-inducing meetings, conference rooms at Open Market have no chairs. Employees take vacations when they want. When they come to work, they're free to bring their dogs along and let them run around the office. There are no job titles; everyone has three responsibilities: Do the work you were hired to do; do something for personal development; and handle an office task.

View from Open Market's office window

http://www.openmarket.com/boscam/index.html

Three personalities are behind Open Market:

- Shikhar Ghosh, who is the chief executive officer, was CEO of Appex Corporation, a company that established a payment system, clearinghouse, and call validation infrastructure for the cellular telephone industry. Prior to Appex, Ghosh was a partner in the Boston Consulting Group and graduated with an MBA from the Harvard Business School.

- David Gifford, a cofounder of Open Market and the company's chief scientific officer, is a professor of computer science at the Massachusetts Institute of Technology. He is an expert on managing information systems on open networks and on payment systems. He is also an advisor to government agencies involved in the National Information Infrastructure.

- The third member of the Open Market management team is chief technology officer Lawrence Stewart. He has seventeen years' experience in computer systems design, internetworking, and multimedia. He has held positions at Xerox's Palo Alto Research Center, Digital Equipment Corporation's System Research Center, and Cambridge Research Laboratories.

There are no job titles; everyone has three responsiblities: do the work you were hired to do; do something for personal development; and handle an office task.

> **The company warns prospective Internet businesses that if they try to apply traditional, hard-hitting advertising techniques to an interactive medium, they will fall flat on their faces.**

This offbeat corporate culture underlines Open Market's message to businesses who want to do business in cyberspace: It's a whole new world out there.

The company states in its marketing brochure, a 30-by-18-inch four-color poster with wacky cartoon characters illustrating the trials and tribulations of doing businesss through the electronic ether: "A lot of markets don't understand the Internet. They're afraid of it. At night they dream about millions of geeks coming to take away their children."

The notion that the Internet is hostile to business comes from the widely reported experiences of a few businesses that decided it was okay to be a bull in a china shop, the company contends. All they saw was a new pipeline, not a new world.

The brochure explains that people on the Internet don't like to get intrusive commercial messages any more than you like hearing from life insurance salespeople at dinnertime. But if you have something they want, it's a different story.

"The Internet is a two-way street," Open Market says. "One of its biggest advantages over other media is that when you talk to your customers, they talk back. Feedback is part of the deal. It's free. It's interactive. It's powerful."

The company warns prospective Internet businesses that if they try to apply traditional, hard-hitting advertising techniques to an interactive medium, they will fall flat on their faces. "This is not an audience with glazed-over eyeballs," Open Market says. "This is an audience that is paying attention. Ad messages need to be a lot more varied, subtle, and content-rich."

Cyberspace is about what people want, the company continues. Computers will be doing more and more of the work. People will tell them what to look for, and their computers will send out search parties to comb the planet.

The company explains that until now, most of the control has been in the hands of the middlemen—agencies, networks, and other media outlets—who tell a business how and when to talk to its audience and how much it will cost every time you want to do it.

That isn't the case with the Internet, because the Internet is an open system. "The people who have the power are the consumers and the content providers, not the people laying the pipe," the company says. "The Internet is a flat competitive landscape—the technology is open to everyone.

"The Internet isn't just another distribution channel," Open Market declares. "It is a new medium. One that will change the way businesses do business. Electronic commerce is a whole new world. With new product creation techniques, new models for customer service, new everything."

Going out to the warehouse, dusting off some old products, digitizing them, and uploading will be a big mistake, the company says. Everything a company has been doing has to be rethought before it enters Internet commerce.

That's where Open Market enters the picture. It sees its role as helping businesses do what they do best and shaping that into the kinds of products that make sense on the new medium.

Like First Virtual, a lot of what Open Market does has to do with creating an atmosphere where buyers and sellers can perform secure transactions on the Internet. Although Open Market isn't minting digital money, it, like First Virtual, is pioneering techniques to make the Internet a safe place for cyberbucks, for if you can't perform a simple credit card transaction on the Net, how are you going to pass silicon simoleons?

According to Open Market, it can offer businesses a highly secure environment for on-line transaction processing. That includes real-time authentication and authorization of credit card transactions and on-line communication for resolving disputes. And it can be done for businesses on untrusted networks as well as secure ones. "You could call it a worldwide infrastructure for electronic commerce," the company says. "Or you could just say that we figured out all the important stuff."

> **Going out to the warehouse, dusting off some old products, digitizing them and uploading will be a big mistake, the company says. Everything a company has been doing has to be rethought before it enters Internet commerce.**

*Open Market's
Home Page*

http://www.openmarket.com/

*Detail of a button on
Open Market's
Home Page*

Connecting Customers and Merchants

Like First Virtual, Open Market recognizes the need for a payment system on the Internet to process small as well as large transactions. It can handle purchases of all sizes. Without this capability, many vendors couldn't afford to support small purchases and would have to charge hefty prices for their information. That approach imposes de facto discrimination on information. Only buyers with deep pockets, like corporations, could afford to buy these merchants' goods.

Open Market allows merchants to sell their wares any way they can think of slicing and dicing them—by a single cut from a CD to a single page from a book. It attempts to connect customers and merchants in a number of ways. It lets customers check their purchases online. This cuts down on service calls and chargebacks.

Open Market also offers its merchants state-of-the-art technology in the implementation of intelligent agents, the computer-based gophers that run errands for users online. These enable customers to discover a merchant's products in the ocean of information occupying the Net.

How Open Market Works

In the Open Market system, content, from electronic catalogs to subscription databases, is held by content or merchant servers. These servers can be centrally or independently managed. The content server is a high-performance Web server—the Open Market WebServer or Open Market Secure WebServer.

Open Market's Secure WebServer, the first Web server software in the industry to support secure Web browsers, features support for two emerging protocols to secure transaction information on the Internet: the Secure Hypertext Transport Protocol (S-HTTP) and a rival technology developed by Netscape, the Secure Sockets Layer (SSL). Both are aimed at securing online transactions from prying eyes through such means as public key encryption.

According to Open Market, S-HTTP is superior among security approaches in its ability to authenticate clients, fully authenticate servers (including checks of revocation lists), support digital signatures (attesting to a message's authenticity), negotiate security levels based on application needs, and provide secure communication through existing corporate firewalls. Firewalls are a hardware and software approach to security that restricts access to a network; all communication—communication from the network to the Internet and from the Internet to the network—is forced to pass through the firewall.

Encryption can provide excellent security against network wiretapping, but still be vulnerable to an electronic break-in to the end user's desktop computer. Open Market goes beyond encryption to protect the business and the network against other kinds of threats by offering

- Flexibility in setting the level and method of security used. Security

> **Encryption can provide excellent security against network wiretapping, but still be vulnerable to an electronic break-in to the end user's desktop computer.**

levels can be set differently for each transaction. A $1 purchase need not have the same security as a $10,000 wire transfer.

- Flexibility in authentication. Methods available include passwords, challenges, public key certificates, or even token devices or smart cards which reside entirely outside the network—protecting against workstation compromise.

- Management of transaction authorizations internally through access control lists and account profiles, or externally, through real-time financial authorization networks.

- Management of fraud through comprehensive statements and audit trails that can be augmented by pattern matching at the payment system.

- Protection of content through document fingerprinting.

When Open Market announced its secure servers in March 1995, CEO Ghosh declared: "These new products allow companies to do business over the World Wide Web on their own terms. They support electronic commerce in a way that transcends mere transactions by supporting ongoing customer relationships and the creation of brand identity."

The secure server products are browser-independent. They are designed to support the entire spectrum of browsers being used by Web crawlers today. "The Open Market WebServers are not derivatives of freeware server implementations," explains Robert Weinberger, Open Market's VP of marketing. "They have been architected from scratch to achieve the highest levels of performance, functionality, and interoperability."

The company says that the servers support in excess of 1,000 simultaneous connections—several times the capacity of any other existing server.

The servers also sport customized access control, which provides total flexibility in determining which users have access to which content under what circumstances. The servers also feature extensive logging facilities, allowing users to completely characterize activity on their site.

The Web servers in the Open Market system send information to the company's Integrated Commerce Service (ICS), which is the meat and potatoes of the system. This service enables businesses to exploit the full potential of the electronic medium on their own terms. Features incorporated in the ICS include real-time credit card authorization and settlement; buyer authentication; customer profiles; subscription services; support for pay-per-page fulfillment services; payment processing; consolidated, up-to-the-minute transaction statements; and customer service and dispute resolution facilities.

The connection between the merchant server and ICS is made with the company's Transaction Link software. This software allows multiple content servers on the system to securely communicate pricing and product information to the ICS. A merchant can set up his own server and get seamless support from the payment server over the Internet through Transaction. As demand warrants, the content servers can be distributed and replicated, without changing the payment system. The system has a variety of fraud control systems and creates an extensive audit trail of all activities.

The payment system can handle any type or size of transaction, from pennies to tens of thousands of dollars. The system will also enable a user to use all forms of payment, whether credit cards, corporate accounts, direct deposit transactions, debit balances, or whatever is acceptable to the seller and buyer. Conceivably, this could include electronic cash. It even supports multiple currencies and nonstandard payment schemes. For example, a business can choose to be paid in service credits or frequent-flier miles.

The system will allow businesses to sell products in various ways. For example, a business could sell books by the page, or sell subscriptions, or sell hard goods that are fulfilled through some other company.

It also is designed to create a single account for any buyer. The system enables a buyer to shop at independent merchants that have Transaction Link software and manage their transactions through a single account. Buyers can set their own spending limits or designate special privileges such as memberships within the account.

Unlike First Virtual, Open Market offers services that go beyond securing transactions for merchants. It also will help merchants manage the content on their servers and deal with customer service.

Managing information in large sites can be difficult and expensive if the service a provider is offering takes advantage of the Internet's ability to handle large amounts of information in different formats. The problem becomes almost unmanageable when the service has interactive areas (chat, forums), multiple levels of membership and access, large numbers of users, constantly changing information (news, special features), and multiple sources of information and editors.

Open Market can help a merchant develop tools for efficiently managing the process of creating new content, tools such as content staging areas, where policies are automatically reviewed and quality checks automatically performed; script management systems, by which scripts, or automatically executed chains of commands, are efficiently generated and managed; content directory and mapping systems to manage multiple levels of content; and enhanced authentication systems, by which single access can be authenticated across multiple sites and multiple accesses can be authenticated within a defined time.

Customer service can be a headache for any business, regardless of size. In many areas of business, such as the credit card and telephone industries, the costs and business impact of customer service often exceed those of other businesses. The cost of fraud in the credit card industry, for example, is a fraction of the cost of answering customer queries and resolving disputes. As the Internet emerges as a major business venue, there's no reason to expect conditions will be any different there. Open Market has developed a system that, it says, will reduce the costs of providing customer service and managing customer queries. The key features of this system are a customer-focused design that enables customers to answer their own questions without talking to an operator; "smart statements" that link customers directly to the pages they looked at while making a transaction—or even to the item they purchased, in the case of information products; feedback forms that are structured to make the process of handling a query efficient; audit trails that track all aspects of transactions; and reporting tools that make it easy for service representatives to get the information they need.

Creating a Market

Just a little more than a week after announcing its secure Web server software, Open Market cut a deal with First Union, the nation's ninth-largest bank holding company, with assets of more than $77 billion, to set up First Union CommunityCommerceSM, the world's first complete, secured electronic marketplace on the Internet.

First Union

http://www.firstunion.com/

First Union has 35,000 merchant banking customers and 10 million customer relationships as potential participants in Community-Commerce, but any merchant or customer with access to the Internet can conduct transactions in First Union's virtual community.

Under the agreement, First Union agreed to use Open Market's Integrated Commerce Environment to do business and maintain customer relationships on the Internet. First Union will use Open Market to provide high-quality customer service, to connect a variety of established payment systems and networks, and to code account information to ensure confidentiality.

At the time of the announcement, Fred Winkler, president of First Union's card products division, stated: "Confidentiality of customer information is critical to the success of electronic banking and, until now, has been the missing link for providing online transactions. After extensive research, we found Open Market, Inc.'s unique, secure payment infrastructure meets our needs and successfully enables financial transactions between merchants and customers. We also believe it will be the foundation for home banking and a full range of online financial transactions."

> "The alliance with Open Market, Inc. will also lay the ground work for full-service home banking through the Internet."
> Fred Winkler

CommunityCommerce is both an electronic marketplace and a bank on the Internet, allowing customers to shop in virtual storefronts and buy and bank online. It allows access to a myriad of information and business development opportunities for small to mid-sized companies.

Open Market's electronic transaction infrastructure enables merchants to save intermediary and inventory charges, while customers enjoy quicker delivery, discounts, and frequent-buyer reward programs. Through the Internet, customers can make more informed buys, quickly and conveniently. For example, concert tickets can be selected after looking at the auditorium's seating chart, or airline tickets can be bought after thoroughly comparing costs and schedules.

At the beginning of 1995, First Union began offering a robust financial services home page on the Internet and began its development of CyberbankingSM, the company's vision of the future delivery of financial services online.

"Traditionally banks have been the center of commerce in local communities," explained Jonathan Guerster, director of financial services for Open Market. "First Union's CommunityCommerce is a natural extension of that role as the Internet grows and becomes a more viable means of transacting business."

Winkler added: "The alliance with Open Market, Inc. will also lay the ground work for full-service home banking through the Internet. The same technology that will enable us to securely process credit card transactions will be enhanced to provide access to account statements from the desktop, allow customers to transfer funds, purchase investment products, and apply for mortgage loans online over the Internet."

Open Market has also brought into its client stable Bank One, which has assets of $87.8 billion. The bank plans to use Open Market to connect its business partners to customers via the Internet.

Bank One Electronic Commerce Home Page

http://www.eft.bankone.com/

In the first application of this new relationship, Bank One, Open Market, and RoweCom, a service provider to libraries and research institutions, have become partners in a system that streamlines the relationship between libraries and publishers. Using an electronic catalog on a PC, libraries can indicate which publications they would like to buy. This information is sent via the Internet to a server, which automatically forwards the order information to the appropriate publisher and the payment information to Bank One. "This simple electronic process handles ordering and payment in one step instead of many, saving more time and money than with traditional procurement methods," explains Richard R. Rowe, president and CEO of RoweCom. Client service will be handled through the same Internet channels. This service is especially important to libraries that need to reduce costs while maintaining inventory.

RoweCom web site

http://www.rowe.com/

Welcome to RoweCom

TOP NEWS

07 July 1995: Subscribe96 Demonstration Software is now available for downloadin

06 July 1995 RoweCom is pleased to be able to provide a September Trial Offer of Subscribe96 Library Edition. This trial will enable libraries to use the software , with payments to a selected group of STM publishers via the payment/service network es Bank One of Columbus Ohio.

In the next phase of the partnership, Bank One and Open Market will incorporate other applications for conducting financial service transactions over the Internet and World Wide Web, such as electronic bill payment and online applications for financial service products.

On the heels of the First Union deal, Open Market inked a pact with Lexis-Nexis, which has 5,800 sources of news and business information online, including 2,400 in full text, to set up Small Business Advisor on the Internet (http://www.directory.net/lexis-nexis/).

Designed for small-business owners, telecommuters, and others involved with home-based businesses, the Small Business Advisor offers a collection of more than 5,000 articles, which are updated daily, Monday through Friday, for Internet users to browse and purchase. The service also incorporates links with relevant newsgroups and news forums on other Web sites, extending the role of Lexis-Nexis, which is based in Dayton, Ohio, from an information provider to an interactive resource.

Open Market's project with Lexis-Nexis, which is a division of $5 billion publishing giant Reed Elsevier, is an example of how the Internet can be used by a large information provider to reach an audience that would have found it prohibitively expensive to subscribe to its services.

"Small Business Advisor is designed to give Internet users the kind of high-quality information service that Lexis-Nexis represents to large corporations and firms today," said Joseph C. Rhyne, vice president and general manager of business information services for Lexis-Nexis. "Through our relationship with Open Market, we can now offer 24-hour, easy-to-use access to Lexis-Nexis information with critical commerce services such as flexible pricing and billing that small businesses want."

Users can browse at no charge for as long as they like. Using the Small Business Advisor's extensive, graphical interface, users can easily locate articles by choosing from topic areas or by searching the entire database with key words and are then given the headline and a short preview of relevant articles with the option to purchase the full text.

Through Open Market's Integrated Commerce Environment, users are authenticated and can then purchase articles "while they wait" via real-time credit card authorization. Purchased articles can then be viewed online, printed, or downloaded to the user's hard drive. Prices start at $1.95 per article, with weekly specials. Unlike other services, this pricing accommodates the needs of occasional users for whom transactional pricing makes more sense than monthly subscription rates.

Open Market's Integrated Commerce Environment includes facilities to connect with a variety of established payment systems and networks for safe credit card purchases, encryption of customer account information to ensure confidentiality, and customer profiling to enable high-quality service.

Also, users can review purchasing history and gain access to online customer service with Open Market's Smart Statements. "I normally leaf through more than thirty publications a month to find the type of information Small Business Advisor provides quickly and easily," said Elizabeth Waldron, a small-business consultant with the Albuquerque-based Paul Tulenko Institute. "Information within Small Business Advisor comes from an enormous database and is pre-screened and thoughtfully organized, creating an invaluable up-to-date time-efficient service."

Open Market gained even more credibility for its approach to Internet commerce when three of the world's leading media companies—Tribune Company, Advance Publications, and Time, Inc.—announced they had selected the company to provide software and services to support their Internet excursions.

The Tribune Company publishes six daily newspapers and owns and operates eight television and six radio stations. Advance Publications owns Condé Nast and Random House. Advance, Tribune, and Time are all on Open Market's board of directors.

Time, America's largest magazine publisher and a major book publisher, will use Open Market's Integrated Commerce Environment to support registration and future commercial activities for its Internet site,

Open Market gained even more credibility for its approach to Internet commerce when three of the world's leading media companies—Tribune Company, Advance Publications and Time, Inc— announced they had selected the company to provide software and services to support their Internet excursions.

Presenting Digital Cash

one of the most stunning on the World Wide Web (http://pathfinder.com), where it provides text, photos, graphics, audio and video from *Time, Sports Illustrated, Money, People, Vibe, Entertainment Weekly, Sunset, Southern Living,* Time Life, Warner Books, Little Brown, and other Time Warner divisions.

Time, Inc.'s World Wide Web Home Page

http://www.pathfinder.com/time/timehomepage.html

A page welcoming you to Pathfinder

http://www.pathfinder.com/pathfinder/welcome.html

"After an extensive review, we were convinced that Open Market had the ability to provide first-class support for our active World Wide Web site," said Bruce Judson, general manager of Time, Inc., New Media.

Condé Nast also has a Web site tied to Open Market (http://www.cntraveler.com/). Called Condé Nast Traveler Online, the site makes sophisticated travel information available in an up-to-date, interactive format. Condé Nast says it chose Open Market to develop its entire online presence because Open Market could build a unique infrastructure for publishing in the electronic age.

One feature of the Condé Nast site is the Great Escapes database, a guide to 1,000 accommodations on 250 island and beach destinations worldwide, including information on hotels and resorts, attractions, updated weather forecasts, color photographs, and directions to destinations of choice.

To update the Web site information on a daily basis, Open Market developed custom software called Editor's Desk, designed so that editors can write articles at their PC without having to worry about formatting them for the Web. This technology dynamically generates articles for the Web and creates an "instantaneous publishing" environment, where information can be reused.

Open Market's advanced WebServer was chosen for its ability to allow an unlimited number of would-be vacationers to browse and interact with the site simultaneously. Open Market's server offers Condé Nast the ability to grow and expand the site to include multimedia content with services. Future plans include leveraging Open Market's secure transaction infrastructure for making reservations and purchases online.

After the Time-Tribune deal, Open Market announced it was teaming up with the Copyright Clearance Center in Danvers, Massachusetts, to bring the power of the Internet to a nettlesome area for payment transactions: copyright law. Finding the owner of a copyright can be a major research project. And dealing with thousands of people seeking permission to photocopy articles or book chapters for classes or training sessions can be a headache for a copyright holder.

Copyright Clearance Center Online

http://www.openmarket.com/copyright/

The Copyright Clearance Center is a not-for-profit corporation formed in 1978 by authors, publishers, and users at the suggestion of Congress. Over 5,000 corporations and their subsidiaries (including 80 percent of the Fortune 100 companies) as well as law firms, document suppliers, libraries, universities, copy shops, and bookstores use the Copyright Clearance Center's photocopy licensing services to facilitate compliance with U.S. copyright law.

Using technology from Open Market, the Copyright Clearance Center has created an online rights clearance system that streamlines the copyright permissions process for more than 1.7 million titles from over 9,000 participating publishers.

The new World Wide Web service on the Internet is called "CCC Online" and is the first of its kind in the world. CCC Online eliminates the two major obstacles people encounter when seeking permission to photocopy copyrighted materials: the time and cost of locating individual rights holders. Using CCC Online, customers know instantly whether they have permission and how much it will cost. Customers can easily search CCC's extensive catalogs for specific titles, find out the total fee, file electronically for permission to copy, and set up payment online. The service is available seven days a week, twenty-four hours a day.

According to Copyright Clearance Center president and CEO Joseph Alen: "This new offering dramatically eases the process of obtaining permissions, which in turn increases the likelihood for publishers and authors to be reimbursed for copying their material. Using Open Market's solution for electronic commerce, CCC Online encourages ongoing customer relationships that will leverage the information access resources of the Internet for publishers and authors."

The extensive Copyright Clearance Center catalogs are searchable by anyone with access to the World Wide Web. CCC Online supports electronic reporting by customers seeking only photocopy permissions, such as academic libraries, document suppliers, and small corporations, as well as customers seeking permissions to create academic coursepacks.

11 MORE PLAYERS

The attraction of digital cash grows every day. Fifteen years ago, when David Chaum spoke of cyberbucks, he was greeted with smirks and the figurative pat on the head reserved by self-proclaimed realists for visionaries whose sight extends beyond today's horizon. Now, as the advent of silicon smackeroos approaches, a number of companies are positioning themselves to be players in the market. These companies include banks, credit card firms, and technology enterprises.

First Union

First Union, which chose Open Market to set up its pioneer electronic marketplace, has aggressively pursued business on the Internet. It offers to Web crawlers a basket of bank services as well as a minimall.

First Union teamed up with MCI in November 1995 in an eighteen-month pilot project that involves performing secure transactions at the bank's Web site. The bank distributes a Web browser, a Netscape browser customized by MCI, to customers through an 800 number. Other distribution channels include First Union branches, direct mail, and client relationship managers.

The software, valued at more than $75, is given away free of charge through First Union, but customers have to pay for Internet access through MCI. The long-distance carrier charges a base rate of $9.95 a month, which includes five hours of Internet access, and $2.50 for each additional hour of access.

Vinton Cerf, sometimes called the father of the Internet and now an MCI senior vice president, has called the MCI–First Union pilot a "giant step" toward making the Internet as powerful a banking tool in the 1990s as the automated teller machine became in the 1980s. Before that can happen, though, some way is going to be needed to dispense cash through the medium.

First Union offers a line of home banking services through Direct Bank, a virtual financial institution at its Web site. Through Direct Bank consumers can apply for auto loans, home mortgages, and credit cards, set up a savings account, buy home or auto insurance, and transfer funds electronically.

Bank One

Bank One has been working with CommerceNetBanc, a joint effort by the government and a consortium of large companies to develop techniques for secure Internet-based commerce, and Open Market to set up secure sites for business to be transacted on the Internet.

In addition, it has been in the forefront of spreading the gospel about electronic payments on the Internet through EDI. Many consumers believe they're paying their bills electronically because they pay them by phone, computer, or automatic transfer. But that isn't the case. About 65 percent of all payments made through home banking service providers are made by check. Relatively few companies accept electronic payments. Most payments are sent in the form of a check-and-list—one check with a listing of all the consumers paying their bills. These checks-and-lists typically require special handling in the receiving company's remittance processing department because they have inconsistent formats, reconciliation, keying errors, or inadequate information. This often results in delays in posting to the customer's account. Bank One can deliver those payments electronically to billers in any format they specify.

One area Bank One is very interested in is authentication, or verifying that the parties in a transaction are who they say they are. As more and more buying and selling takes place on the Net, the need for these services will be enormous. If public key encryption takes hold, somebody has to be the holder of the public keys and guarantee their validity. Bank One thinks that someone will be someone like itself.

Visa

McDonald's and Coca-Cola are the early enlistees in Visa's anticipated plan to introduce digital cash at the 1996 Olympic games in Atlanta.

Visa is also part of the consortium, including MasterCard and Europay International, developing an international standard for smart cards. One area the consortium is looking into is creating an international electronic purse specification that will ensure global acceptance and interoperability of stored-value cards and the digital cash in them.

VISA's Home Page

http://www.visa.com/

One World, One Currency

VISA

Products

The Future is Visa

Financial Tips for Consumers

News from Visa

Visa/PLUS International ATM Locator Guide

Lost Card Information

Special Events & Promotions

MasterCard

Late in 1995, MasterCard will begin an electronic cash pilot program in Canberra, Australia, called MasterCard Cash. MasterCard's hopes for the pilot's success have been buoyed by market research it's done Down Under and Stateside.

"By design, MasterCard's stored-value application meets the needs of consumers—globally," says Diane Wetherington, senior vice president of chip card business and marketing for MasterCard. "Our chip card strategy is very focused: that is, to provide our worldwide membership with new global payment vehicles to build relationships with consumers. In planning for our pilot in Australia, MasterCard recognized the similarities between Australian and U.S. consumers. The results support our interest to apply the pilot results from Australia to the U.S. and beyond."

If public key encryption takes hold, somebody has to be the holder of the public keys and guarantee their validity. Bank One thinks that someone will be someone like itself.

Presenting Digital Cash

MasterCard International Pointers

http://www.mastercard.com/

One startling discovery revealed in MasterCard's research was that consumers were so eager for some sort of stored-value vehicle that they would switch banks to obtain the service. In the United States, 60 percent of respondents said they would switch banks to obtain the stored-value feature; 55 percent of Australians surveyed would be willing to do the same.

The quantitative consumer concept studies, conducted with almost 2,000 people in the United States and Australia, revealed other significant trends as well. First, consumer interest in stored value is largely based on a consumer's lifestyle and does not vary significantly across different demographic groups.

Second, consumers consider the stored-value application most valuable as complementary to coins and cash, and not as a cash replacement. Finally, mass market outlets such as convenience stores and gasoline stations are where stored value is likely to have the greatest impact.

The clearest result of the studies was the striking similarity in response to the stored-value application by American and Australian participants. More than half of all Australians expressed positive interest in the stored-value concept, while 60 percent of Americans said the same. Other similarities crossed borders as well:

When given a choice between having a stored-value feature linked to an existing ATM/debit or credit account or having a stand-alone card, consumers overwhelmingly chose the ATM/debit or credit option. In the U.S. 84 percent (76 percent of Australians) wanted a stored-value application on an ATM/debit or credit card, while 16 percent (24 percent of Australians) wanted a stand-alone cash card.

The primary merchant outlets were consistent across the U.S. and Australia, the top three being gasoline stations, supermarkets, and convenience stores. Secondary merchant locations differed by geography. Australia: fast food, newsstands, pay phones, and public transportation; U.S.: drugstores, department/discount stores, family restaurants, and vending.

The ATM was the preferred method for loading in both countries: 52 percent of Americans would use ATMs most often to load value; 46 percent of Australians would do so. Another top location included the checkout at the point of sale.

Americans prefer to carry an average minimum of $100 and an average maximum of $300 on their cards; Australian respondents said $50 as the average minimum, $250 as the average maximum. The overall range of value requested was surprisingly broad—as low as $25 to as high as $1,000. Key benefits included: convenience (91 percent of Americans, 93 percent of Australians); usefulness for small-value purchases (44 percent of Americans, 49 percent of Australians); universal acceptance (28 percent of Americans, 33 percent of Australians); and safety (39 percent of Americans, 40 percent of Australians).

MasterCard has also joined forces with Visa to create specifications for maintaining secure transactions on open systems like the Internet. The security specification is expected to be ready in 1995, and the companies say secure transactions will be taking place on the Internet in 1996. The specification supported by MasterCard and Visa will be open and available to everyone. This standard will provide payment security for all bankcard transactions; other security protocols can be used to protect personal data. The new standard will also facilitate deployment of personal computer software to incorporate payment security applications.

The specifications supported by the associations will call for the use of extensive encryption capabilities based on RSA Data Security to protect card transactions on the Internet and other networks. And MasterCard and Visa anticipate that purchases and payments performed on open networks such as the Internet will function much like other bankcard purchases.

MasterCard International: Visionarium

http://www.mastercard.com/Vision/encrypt.htm

Financial Services Technology Consortium

A recent arrival in the electronic payments arena, the Financial Services Technology Consortium, is floating a plan to create an electronic check system for the Internet, using smart cards. A pilot program is expected to be launched in 1996 by the consortium, which consists of CitiCorp, BankAmerica, IBM, and Sun Microsystems.

The check system has some of the flavor of electronic cash. Users who request merchandise from an Internet vendor would insert their smart card into their computer. An invoice and payment would be enclosed in a protective electronic envelope. The envelope would be sent to the merchant through the Internet. The merchant would forward the electronic check to his bank for clearance over existing banking networks. Foreign currencies and certified checks could also be handled by the system.

CheckFree

CheckFree has started marketing an electronic "wallet," but it doesn't have anything to do with cash. It essentially allows a user to perform a secured credit card transaction on the Internet. What's interesting about the approach is Checkfree's choice of terminology—calling its software a wallet rather than a credit card.

The CheckFree Wallet consists of a "consumer purchase center" built into a user's Web browser. The user loads her purchase information onto the purchase center, where it is stored under heavy encryption for security. Because the user's shipping and credit card information is already stored in the CheckFree Wallet, she doesn't have to fill out lengthy order forms every time she makes a purchase.

The wallet has been integrated into the Spyglass Enhanced Mosaic browser software. Soon, it will be available in more places and will allow for purchases by check as well as credit card.

*The CheckFree
Home Page*

http://www.checkfree.com/

CheckFree
Payment Services

CheckFree <u>frees you from checks</u> and secures the way you <u>purchase</u> on the Internet. If you haven't tried it, we're so sure you'll like it, we'll give you the <u>software for free!</u>

CheckFree is <u>safe</u>, secure, and easy to use. And with our new <u>low price</u>, it's even cheaper than using stamps.

But don't just take OUR word for it. <u>Listen</u> to what some CheckFree users have to say!

 And here's <u>What's New</u> at CheckFree!

Paying   Buying   Caring for You   Find it Here   The Suits   Our Company

CheckFree Explains Its System

http://www.mc2-csr.com/vmall/checkfree/v20/wprocess.html

CommerceNet

CommerceNet is an umbrella organization created to promote electronic commerce on the Internet. It is a step toward a de facto national information infrastructure capable of linking up with other electronic commerce projects in places such as Boston, Austin, and the University of Illinois. Potentially, a CommerceNet-like infrastructure could support other national efforts in education, health care, and digital libraries.

> **CommerceNet sees this new Internet-based electronic marketplace supporting all business services that normally depend on paper-based transactions.**

CommerceNet's charter, posted on its World Wide Web page, outlines the functions of the organization:

- To operate an Internet-based World Wide Web server containing information that will facilitate an open electronic marketplace for business-to-business transactions.

- To accelerate the mainstream application of electronic commerce on the Internet through pilot programs on such subjects as transaction security; payment services; electronic catalogs; Internet EDI; engineering data transfer; and design-to-manufacturing integration.

- To enhance existing Internet services and applications and stimulate the development of new services.

- To encourage broad participation from small, medium-sized, and large companies and offer outreach programs to educate organizations about the resources and benefits available from CommerceNet.

- To serve as a common information infrastructure for northern California and coordinate with national and international infrastructure projects.

CommerceNet's goal is to stimulate the growth of a communications infrastructure that will be easy to use, oriented to commercial use, and ready to expand rapidly. The net result for businesses will be lower operating costs and faster dissemination of technological advances and their practical applications.

CommerceNet sees this new Internet-based electronic marketplace supporting all business services that normally depend on paper-based transactions. Buyers will browse multimedia catalogs, solicit bids, and place orders. Sellers will respond to bids, schedule production, and coordinate deliveries. A wide array of third-party value-added information services will spring up to bring buyers and sellers together. These services will include specialized directories, broker and referral services, vendor certification and credit reporting, network notaries and repositories, and financial and transportation services. Although many of these transactions and services are already available electronically, they require dedicated lines or the making of prior arrangements. The use of an Internet-based infrastructure reduces the cost and lead time of participating in electronic commerce, and makes it practical for both small and large businesses.

More Players

CommerceNet also provides a forum for industry leaders to discuss issues and deploy pilot applications, and from these to define standards and best business practices for using the Internet for commerce. Through these efforts, CommerceNet will help this emerging industry evolve common standards and practices so that users will see a seamless web of resources.

CommerceNet also helps companies participate in Internet-based electronic commerce. It provides

- Affordable, high-quality Internet connectivity using a variety of options, including high-speed connections.

- Online directories of CommerceNet members and subscribers, and other electronic commerce initiatives.

- Online access to software tools for information providers that enable them easily to set up their own Internet presence.

- Security mechanisms, including authentication and encryption, supported within applications, including RSA public key cryptography and public key certification.

- Outreach focused on making it easy for small businesses to participate as Internet information providers.

CommerceNet works on new developments at the Stanford Center for Information Technology. Future technologies being explored include intelligent shopping agents that can search through catalogs and negotiate deals; collaboration tools for distributed work teams that support both real-time interaction and videomail; natural-language search-and-retrieval techniques for large distributed-information bases; and format translation services that enable engineering organizations to exchange product data even when they adhere to different standards.

CommerceNet believes that the majority of companies and organizations in the United States may conduct business via the Internet in five years.

CommerceNet's Home Page

http://www.commerce.net/

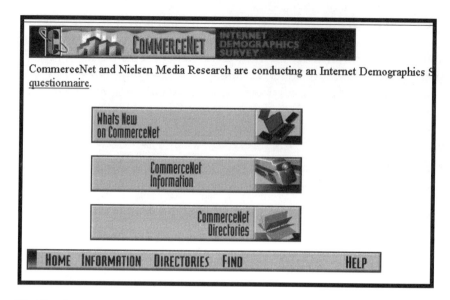

NetCash

NetCash is an electronic currency being developed at the Information Sciences Institute (ISI) at the University of Southern California.

NetCash is designed to facilitate anonymous electronic payments over an unsecure network without requiring tamper-proof hardware. It aims to secure transactions in an environment where attempts at illegal creation, copying, and reuse of electronic currency are likely. In order to protect the privacy of parties to a transaction, NetCash implements financial instruments that prevent traceability and preserve users' anonymity.

NetCash is designed to be scalable. That means the electronic currency can be accepted across multiple administrative domains. Currency issued by a currency server is backed by account balances registered with NetCheque, an electronic check payment system also based at USC, to the currency server itself. NetCash currency servers also use the NetCheque system to clear payments across servers and to convert electronic currency into debits against and credits to customer and merchant accounts. Though payments using NetCheque originate from named accounts, with NetCash the account balances are registered in the name of the currency server, not the end user.

The NetCheque system was developed by a team led by ISI computer scientist Clifford Neuman. It combines existing computer security technology initially developed at the Massachusetts Institute of Technology, with distributed authorization and accounting technology developed at ISI, to represent and process electronic financial instruments, similar to checks, that one can use to pay for goods and services over the Internet.

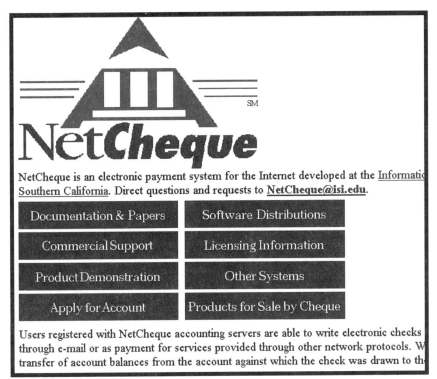

The NetCheque Network Payment System

http://nii-server.isi.edu:80/info/netcheque/

NetCheque works much like a conventional paper check. An account holder issues an electronic document that includes the name of the payer, the name of the financial institution, the payer's account number, the name of the payee, and the amount of the check. Most of the information is in uncoded form.

Like a paper check, a NetCheque bears the electronic equivalent of a signature—a code group that authenticates the check as coming from the owner of the account. And, again like a paper check, a NetCheque needs to be endorsed by the payee, using another electronic signature, before the check can be paid.

Finally, through NetCheque, properly signed and endorsed checks can be electronically exchanged between financial institutions through electronic clearinghouses, with the institutions using these endorsed checks as tender to settle accounts.

Presenting Digital Cash

NetCheque secures its transactions through Kerberos software—developed by Neuman and others at MIT—which is now widely used to protect network computing services from break-ins and theft of services.

Named after the three-headed watchdog of Greek mythology, Kerberos is designed to allow access without passwords or other compromising information being sent over network links. In a series of coded back-and-forth messages between two corresponding computers, Kerberos creates a coded data packet called a ticket that can be presented to a third computer to securely identify the user.

The Kerberos Network Authentication

http://nii.isi.edu/info/kerberos

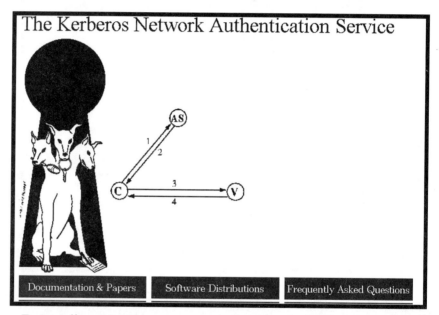

Essentially, a NetCheque is a specialized kind of ticket created by the Kerberos system. A user's digital signature is used to create one ticket—a check—which the payee's digital endorsement transforms into another—an order to a bank computer for fund transfer. Subsequent endorsers add successive layers of information to the tickets, precisely as a large number of banks might wind up stamping the back of a check along its journey through the current system.

The NetCheque software is available free of charge for personal, non commercial, or limited commercial use. For commercial use or integration with commercial products, USC will grant nonexclusive licenses on generous terms.

Netscape

Netscape makes one of the most popular browsers for the Web. It is building into its browsers the ability to secure transactions on the Internet. Some companies, like MCI, have customized Netscape's browser so secure transactions can be made through it.

Netscape was founded in April 1994 by Dr. James H. Clark, founder of Silicon Graphics, a Fortune 500 computer-systems company; and Marc Andreessen, creator of the NCSA Mosaic software for the Internet. Mosaic is the software that brought the graphic interface to the Net.

Netscape also makes Secure Courier, the first open, cross-platform protocol to create a secure digital envelope for financial data on the Internet. Among the protocol's supporters are Intuit and MasterCard International, who plan to use the protocol for securing online credit card, debit card, charge card, and microfinancial transactions.

The new open protocol builds on existing industry standard protocols including the Secure Sockets Layer (SSL), which is built into Netscape's browsers. Secure Courier supports the MasterCard and Visa security specification for bank card purchases on open networks

Compatible across Unix, Windows, and Macintosh operating environments, Secure Courier increases security for commerce on the Internet by encrypting a consumer's financial information all the way from his or her PC to the financial institution. In addition, Secure Courier enables consumer authentication for merchants. While secure-channel protocols such as SSL encrypt data passing along the network between a client system and a server, Secure Courier delivers the additional security of keeping financial data encrypted—or in a "secure digital envelope"—when they arrive at a merchant's server or at other intermediate points on the Net. This means that the data remain "wrapped" or protected at any site at which they stop. The Secure Courier protocol decreases the risk of consumer and merchant fraud, enables global payment security, and reduces merchant costs.

Secure Courier is a key part of Netscape's complete open payment system for online transactions, which consists of three phases:

- Enabling reduced-cost commerce across the Internet by using SSL to accept orders on the Net, and authorizing and settling transactions offline. Netscape implemented this phase during the first quarter of 1995.

- Building the infrastructure for PC-to-bank secure payments. This phase includes the publishing of the Secure Courier protocol specification and the licensing of Secure Courier to partners. This phase is designed to establish the infrastructure for increased security through a digital certificate and end-to-end encryption of financial transactions.

- Delivering products that include Secure Courier for end-to-end secure electronic payments by the end of 1995. This phase will include the deployment of commercially supported software for consumers, merchants, and financial service providers. With this software, financial transactions can be accepted, authorized, and cleared over the Internet with an increased level of security.

Smart Cash

EPS, based in Wilmington, Delaware, has been planning for years to introduce a stored-value card, which would function rather like the prepaid telephone cards being introduced into the market. The EPS card, however, would be "rechargeable" with cash, and would be designed for use in many more locations than telephone cards.

EPS is part of a consortium of more than a dozen companies developing a standard technology for the cards.

Owners of the MAC bankcard network will issue stored-value cards to several thousand consumers in an area of suburban Delaware sometime in mid-1996.

A spinoff of EPS called Smart Cash is also working on stored-value technology. Smart Cash is jointly owned by MasterCard International; two other card-payment technology companies, VeriFone and Gemplus SCA; and at least eleven banks, including CoreStates Financial, PNC Bank, KeyCorp, Bank One, and National City, which together also own EPS.

Telequip

Telequip, a $6 million company based in Hollis, New Hampshire, makes a pocket-sized personal encryption card called Crypta Plus. The card plugs into a slot in a personal computer and gives a user access to his own unique encryption keys. Once the card is unlocked with a personal identification number, it can be used to send encrypted information, including financial transfer information, over the phone via modem. The card can also be used to store electronic cash. Telequip is also working with the Financial Services Technology Consortium on its electronic checkbook.

NetMarket

NetMarket's claim to fame is that it was the first company to offer automatically encrypted Mosaic sessions for credit card transactions over the Internet. NetMarket performed the demonstration in August 1994. It was an important step in advancing the development of electronic commerce on the Net. It filled in a pothole in the Information Superhighway that was slowing down business development of the Net, and cleared the pavement for the development of payment vehicles, such as digital cash, for completing transactions in cyberspace.

NetMarket, a Delaware corporation based in Cambridge, Massachusetts, connected to the Internet in March 1994 and signed its first storefront, Noteworthy Music, in July of that year. Noteworthy, which offers more than 17,000 deep-discounted CDs, accepted an encrypted transaction from Phil Brandenburger, a system programmer for the Wharton School of Business at the University of Pennsylvania.

NetMarket made use of the public key encryption technology called Pretty Good Privacy (PGP), which was developed by Phil Zimmerman and made available without charge for personal use through licensing by the Massachusetts Institute of Technology (MIT). NetMarket purchased a commercial license for PGP from Phoenix, Arizona–based ViaCrypt and ported the program for use on NetMarket's Unix-based computers.

Three months after its network milestone, NetMarket was acquired by CUC International, a billion-dollar company on the New York Stock Exchange. Ironically, CUC cut its teeth in electronic home shopping fifteen years earlier but had to revise its business plans when projections for PC penetration of the home failed to become a reality. In the interim it has concentrated its efforts on membership-based discount services. About 30 million members buy discounted products from CUC.

Now that PC penetration of the home is approaching the levels predicted fifteen years ago, CUC is getting back into the electronic home market. Along with its Internet efforts with NetMarket, it is working with a range of industry leaders, such as CompuServe, America Online, Time Warner, Dow Jones, Delphi, GEnie, Prodigy, AT&T, and many others. Says CUC CEO Walter A. Forbes, "We have no idea what the winning platform for shoppers is going to be and therefore we're trying them all."

According to Roger Lee, president of NetMarket, the company's role now is primarily as a consultancy. In addition, the company has worked on CUC's mammoth Internet membership club, Shopper's Advantage, which will offer 250,000 products to Net surfers. Initially, Shopper's Advantage will use an offline system to clear credit card transactions. Members will set up their accounts by phone and will be issued a membership number to order products online. The order will then be completed offline, where the member's number will be matched to his credit card information. Secure Web server technologies like Netscape, Spyglass, and Open Market will be supported in the future, Lee says.

12 GLOBAL IMPLICATIONS

The implications of digital cash for global society are enormous. Every pillar of society—governments, corporations, financial institutions—will be affected by it. Every citizen's daily life will change because of it. Here are some ways e-cash may shake things up.

Government

Since the Great Depression, the state has been intimately involved with the economy. Electronic cash could wedge a pry bar between the two, a trend that appears to be in progress even without the help of digital dollars.

The 1990s have been marked by several currency and money-flow crises that suggest governments and central banks don't have the control over the world's money flow that they used to have. Electronic cash isn't going to make their job any easier. Cyberbucks can wing their way between countries at chrome-burning speeds without a trace, which makes it difficult for governments to control and tax the stuff.

How hard will e-cash be to control? Money is now created through the fractional reserve system. Commercial banks have the exclusive right to lend out more money than they have on deposit. If e-cash minters could do the same thing with cyberbucks, institutions outside the control of the central banks would be creating money—just as the commercial banks do now. This could lead to financial chaos. Instead of having an institution like the Federal Reserve controlling the money supply, you could have hundreds of entities deciding how much money they wanted to put into circulation today. For government regulators, that would be a scene from an Edvard Munch painting.

One way to address the money supply problem is to require e-cash minters to back their cyberbucks with state currency on a one-to-one basis. For each cyberdollar in circulation, the company must hold a dollar in reserve. That would keep control of the money supply with commercial banks, not something relished by e-cash proponents, who see cyberbucks as a means of getting money systems out from under the control of governments and into the open market.

If the introduction of e-cash isn't handled carefully, consumers could be left totally clueless about the system. They could be assaulted with a confusing array of competing currencies. If consumers have a hard time understanding that just because a mutual fund is sold through a bank it isn't protected by the FDIC, how will they understand a bank disbursing Coca-Cola cash?

To add to consumers' comprehension problems, there's the fact that e-cash is stored in a card like a credit card. If someone steals your credit card and goes on a spending spree, you're liable for only $50 of those charges. If someone lifts your e-cash card, its contents are gone forever, just as if someone picked your pocket. What's worse, if you have a few grand stored on your hard drive and it crashes, you could lose it all.

These problems aren't insurmountable, although the more privacy a consumer demands, the harder it will be to restore lost e-cash. If you can't trace an e-buck to someone, then it's hard to restore the money to him when he says its been lost or stolen. But that problem aside, opportunities should open up for insurance companies to write all kinds of insurance on possible losses attributed to digital cash.

"The process, for now, resembles the free-for-all that surrounded the U.S. banking industry in the nineteenth century, until the creation of the Federal Reserve," DigiCash boss David Chaum told a congressional committee in 1995. "Before the Fed, banks circulated their own private

Cyberbucks can wing their way between countries at chrome-burning speeds without a trace, which makes it difficult for governments to control and tax the stuff.

currency and bank checks weren't as widely accepted, since you couldn't trust the solvency of the issuer."

He adds that the same pattern is being repeated in the digital marketplace. "Without clear ground rules, uncertainty will undermine e-cash's usefulness," he says. "What's at stake here? At worst, we'll be left with an inflexible currency that's costly to use, easy for marketers to trace, and hard to trade between individuals; at best, we'll get the digital equivalent of a dollar bill—the benefit of cash without the cost of paper."

Right now, there are no rules on how digital cash should be treated for tax purposes. No one knows where a digital money transaction should be taxed. If a buyer in Rhode Island buys something from a seller in Bombay through a virtual bank in Ohio, whose, if any, sales tax do you pay?

Regulators aren't moving very quickly on the issue; in the minds of some people, most notably the American Bankers Association, this regulatory sluggishness is impeding the development of e-cash.

As far as the states are concerned, though, they would be happy to see e-cash development, or anything else that increases Internet commerce, flailing in a bed of quicksand. Here's why.

A key component of the states' revenue base is the sales tax. But each year, the states lose billions to mail-order sales—according to the U.S. Advisory Commission on Intergovernmental Relations, an estimated $3.3 billion, to be exact.

"And this lost revenue is based on current technology," emphasizes a report from the Nathan Newman Center for Community Economic Research at the University of California–Berkeley. "With the growth of the Internet and online sales, consumer access to a nationwide and worldwide marketplace will expand exponentially. At a push of a button, consumers will have access to the lowest-priced goods nationwide and, with the added bonus of avoiding sales taxes, interstate sales may explode over the Internet leaving state and local government finances in tatters."

Banking

Electronic cash has many profound implications for the banking industry. It also has some more superficial ones, like saving the industry money.

Checks and cash, instruments that require physical handling, are a tremendous expense to institutions that conduct the vast majority of

> "The process, for now, resembles the free-for-all that surrounded the U.S. banking industry in the 19th century, until the creation of the Federal Reserve."
> David Chaum

their business through electronic networks. You can have the fastest computers in the world, but there's still only one way to tally cash: count it by hand. And to get a check from here to there: carry it. The Federal Reserve charters forty-seven planes every night of the week to pick up checks and fly them to a location where they can be tabulated a second time. While trillions of dollars may flow among nations every day at near light speed through electronic funds transfers, the cash that's dropped off at tellers' windows or night depositories, and the checks moving through the Federal Reserve system still have to travel through human hands. When we try to change a sawbuck or cash a check at a strange bank and are rejected, we feel the bank is being rude, but from the bank's point of view the fewer checks it handles and the fewer times a teller has to touch a buck, the greater its bottom line will be.

Having to physically handle checks and cash—collect them, store them, transfer them—is just part of the problem. The physical nature of the vehicles makes them more vulnerable to predators than the electronic instruments a bank works with. Cash can disappear without leaving an audit trail. Checks take days to clear, which gives check forgers time to work their mischief. A forger can deposit a check drawn on the account of a major corporation into a bank account opened under a phony name and close the account and bolt with his ill-gotten gains before the validity of that check is ever questioned by the corporation.

If electronic cash reduced the volume of paper—checks and cash—the banks have to handle, they believe they would see an improvement on their balance sheets.

Electronic cash will also save banks money and change the way they do business through its effects on financial institutions' retail trade.

As wired as banks are today, they still remain grounded in geography. Some leading-edge banks have experimented with computer-based home banking and bank officers armed with laptop computers and modems acting as branches on wheels, but most banking still depends on branches. And branches mean expensive real estate and personnel. ATMs have reduced the need for many branches, but ATMs take up real estate, too, and they are expensive to buy and maintain. Electronic cash has the potential to change all that. Indeed, it has the potential to blow it all away and create a global market that defies local, regional, and national boundaries.

Cash is the last barrier to the wired bank. You can apply for loans and mortgages online. You can pay bills without writing checks. You can set up automatic deposits and withdrawals. But you still have to go to an

ATM to get cash. When that last piece becomes wired, when you can slip your smart card into your phone and perform all your banking functions from home, that's when banking becomes truly wired. And once people become accustomed to banking online, they won't care where that bank is located, or if it even *has* any branches. Swiss banks will still have cachet for some depositors, but now anyone who wants a Swiss bank account can have one without flying to Zurich to open it.

This, of course, savages the retail banking business as we know it today. Branches, ATM sites, and ATMs can be eliminated or reduced. And, once again, financial institutions can sock away the savings.

As peachy as these developments may be for banks, they may be sour apples for people living in poor neighborhoods or undeveloped countries. As people in America's inner-city neighborhoods know too well, when a bank loses its physical roots in a community, all too often it loses its interest in the community, too. Or worse, it deliberately steers money away from the community through redlining.

The prospect of global redlining isn't an attractive one, but digital cash might work against this prospect, too. Since anyone can "mint" digital money, what's to stop communities from creating their own currency, or at least finding someone who will set up a bank down the block to do it for them? In that scenario, instead of digital cash becoming a community wrecker, it could be a community builder.

But electronic cash can not only save money for banks, it can make money, too.

As electronic money becomes more and more popular, the banks can charge fees or royalties for using it. These charges could be very small—but multiplied over millions of transactions they could become a lucrative source of income for a financial institution. In addition, as fond as e-cash proponents are of saying digital dinero is "real," it really isn't. When you take cash out of an ATM, that money stops working for the bank. The value is in your pocket. When you take e-cash out of the bank, the bank can still earn money on the "float," the time money spends in your digital wallet. So if the digital cash sits in your wallet for a day or two, that's a day or two's interest the bank can earn on the money that it couldn't have earned on hard cash.

This scenario gets better for the bank if you make a purchase with a merchant who is part of the the bank's e-cash system. Then the money never stops working for the bank. It collects interest while you have the e-cash. It collects interest while the merchant holds the e-cash. Then the cycle is competed with the money being deposited in the bank. For

Cash is the last barrier to the wired bank. You can apply for loans and mortgages online. You can pay bills without writing checks. You can set up automatic deposits and withdrawals. But you still have to go to an ATM to get cash.

> **Banks believe they're in a cat fight now trying to compete with the financial services industry. What happens when they have to compete with global corporations like Coca-Cola, Pepsi, General Motors and General Electric?**

all the time the money was floating in circulation, the bank was earning income on it, not only interest income, but royalty income from you and the merchant with whom you did business. None of that money could have been made by the bank with hard cash.

If this picture appears a bit too rosy for banks, it is. What financial institutions will discover is that banking is necessary for a modern economy but banks are not. With the arrival of e-cash, banks could find themselves fighting for their lives. Why? If anyone can start minting cyberbucks, large companies may decide to get into the business themselves. Once successful in the digital cash market, these companies might set their sights on other consumer financial services now provided by banks. Banks believe they're in a cat fight now, trying to compete with the financial services industry. What happens when they have to compete with global corporations like Coca-Cola, Pepsi, General Motors, and General Electric?

Banks are already beginning to feel the heat. One insider estimates banks have about five years to introduce viable e-cash products before competitors charge into the market like hussars and carve it up among themselves. Even if the banks manage to swing their bulk into motion on the e-cash issue, they still may fall prey to nonbanks who get into the market early enough to set the standards for digital cash systems. By controlling the standards, these companies may force the banks to accept the role of a node on a network controlled by a nonbank. Or, as one industry insider put it, banks may find themselves buttons on a network run by someone else.

If that happened, it wouldn't be the first time the banking industry found itself muscled out of a market by nonbanks. Twenty years ago, banks owned the credit card transaction business. Now 80 percent of that business is conducted by nonbanks. A similar debacle has occurred in the wholesale banking business, where outfits like General Electric and EDS have taken over the market for transferring payment data to corporations.

Electronic cash can make life a lot easier for a merchant. She won't have to make change anymore. Customers who are short on pocket cash can make that impulse buy without running off to an ATM; they can download the extra money at the shop. She won't have to worry about employees filching from the till. Daily deposits are only a phone call away.

But the greatest impact e-cash will have on a merchant is that it will give him the potential to be a true global player. With digital dough, the smallest shop on Merchant's Row has the capability to tap into the global market. The merchant can carry on financial transactions around the world without transferring bank funds. And with the reach of the Internet, he has the ability to reach a large population of consumers, many of them affluent. Moreover, the merchant doesn't even have to have a storefront—not one that takes up earthly real estate, anyway. He can work out of his home and build his storefront on the Internet.

Crime

Next to losing control of the money supply, nothing worries government regulators more than the implications e-cash may have for crime fighters and Internal Revenue agents.

Money laundering and tax evasion could increase as criminals use untraceable silicon simoleons to stash their loot in offshore banks.

Counterfeiters would have a difficult time forging individual cyberbucks, but that approach would be small-time, anyway. Forgers would look for ways to set up their own "mints" and circulate their own form of digiqueer. Or crack an outfit like DigiCash and mint money that would be indistinguishable from the real thing.

And computer hackers or other criminals could break into an e-cash system and instantly filch the wealth of thousands or even millions of consumers.

While the eyes of high-tech hackers may be widening at the prospect of e-cash, street toughs, once they get a sniff of the stuff, won't be happy. Fewer people will be using ATMs, and those who do will be exiting with a card full of cash that thieves will find worthless to them. Vulnerable populations like the elderly will be able to manage their finances from the safety of their homes. Merchants can make after-hours deposits without walking the streets with large sums of cash and risking an encounter with a bop-and-snatch artist. Electronic cash could be a godsend for gamers of every stripe. States could expand their lottery offerings across the globe. Betting parlors could spring up on the Internet like weeds. Bets and payoffs would be instantaneous. Electronic cash may be in its infancy, but this opportunity has already been tapped. The Internet Online Offshore, run out of the Turks and Caicos Islands, says it will accept all manner of e-money and pay customers 10 percent interest on the balances they leave in an offshore bank the casino owners operate.

With digital dough, the smallest shop on Merchant's Row has the capability to tap into the global market.

Turks & Caicos Islands Online Guide

http://www.digimark.net:80/dundas/turksog/

TURKS & CAICOS ISLANDS
Online Tourist Guide
BEAUTIFUL BY NATURE

Welcome to the Turks & Caicos Islands Online Tourist Guide.

The Turks & Caicos Islands offer you now the antidote to stress. Excitement is waiting dolphins play in the turquoise waters, or ospreys diving for dinner. It is discovering an fish that inhabit the miles of fringing reefs. Excitement is catching an elusive marlin in t shallow banks - and returning them both to their natural habitat.

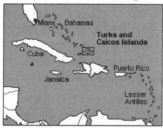

Location of the Turks & Caicos Islands

Presented and Managed by Internet Casinos & Volcano Casino and Amusements, Turk and under the auspices of the W.B. Eugene Family Trust.

Internet Casinos

http://www.casino.org/cc/car-cas.html

Class Issues

A burning issue of the Information Age has been its impact on social class. Some theorists argue that information will merely replace money as the arbiter of class. These souls argue that rather than ushering in a new age of enlightenment, an age in which the individual will be revered, respected, and empowered as never before, an era when the common man will be able to choose his lifestyle without economic constraints, the increased importance of knowledge will create a new class structure just as despicable as the old one. This new structure will consist of those in the know—the information haves—and the clueless—the information have-nots. If this situation develops, will digital cash follow the fault lines? Will there be a class of cyberbuck haves and cyberbuck have-nots?

On its face, this may seem a little absurd. A poor person may be short on money, but most still have some, although it may be a little much to imagine a panhandler shoving his electronic wallet in the face of a passing pedestrian and asking for a handout. If the distribution of electronic money were limited to expensive devices, low-income folks could be locked out of the system, but that flies in the face of two trends. One, the price of devices like PCs is rapidly coming down. By the time digital money gains a market presence, PCs won't cost more than a color TV. Two, access to digital cash won't be limited to computer networks. Stored-value cards will be issued by banks as routinely as they issue ATM cards. This will bolster the proliferation of digital cash, not limit it. Moreover, pinching the market for ether lettuce runs contrary to recent trends in the financial services industry. At a pace that would shame a Formula One racer, companies like Fidelity Investments have developed products to broaden participation in the financial markets. The whole thrust of this activity has been to bring products that had previously been in the sole domain of institutional investors to the public at large. Electronic cash will continue that trend. Now only large institutions can make global payments. With the proliferation of digital cash, everyone will be able to do it.

There are class struggle scenarios, however, in which electronic cash could play a role. As we have seen before, anyone may have the power to issue digital currency—even the government. Millions of people depend on local, state, and federal payments. It wouldn't be hard for these governments to place digital conditions on the e-cash they issue. For example, if a welfare agency disbursed e-cash to a recipient for food,

While the eyes of high-tech hackers may be widening at the prospect of e-cash, street toughs, once they get a sniff of the stuff, won't be happy.

that money could be restricted to food items. Try to spend the money on a bottle of white port or a pack of cigarettes, and the purchase would be annulled at the point of sale. This kind of system smacks of Orwell's *1984*, but it does illustrate how electronic cash could be used by a government to control and intrude into the lives of its less fortunate members.

Privacy

A towering concern for many people involved in the development of digital cash is what impact cyberbucks may have on an individual's privacy. They see e-cash as having the potential for great mischief. To understand why, we need to look at how e-cash will change how we buy information.

How do we obtain our information now? We buy newspapers full of stories we don't read and don't care about. We buy books we skim for nuggets of important information. We watch or listen to broadcast news programs full of reports which have little utility for us.

Why does information have to be packaged this way? Because it's not economical to do it any other way. As you slice the appeal of your information, you reduce the size of your audience. The smaller the audience, the smaller the market. And the smaller the market, the more you have to charge for your information. This narrows your market even further, because you can't sell to everyone in your narrow market who wants your information. You only can sell to people who can afford to buy what you're selling.

There is a way to make your narrow slice of information available to everyone in your narrow market, though. Cut it up into even finer strips and sell it for a small price. The problem there, however, is that as you make more sales at smaller prices, the cost of transacting that sale becomes a substantial part of the cost of your product. Indeed, in today's world those transaction costs make such sales prohibitive. That's why merchants moan when somone tries to buy something costing less than $10 with a credit card. The cost of that transaction is greater than the profit the merchant is making on the item.

How can those transaction costs be brought into line? Through digital dough—because, as Mondex CEO Tim Jones is fond of saying, the cost of transactions with e-cash shrinks to insignificance.

Global Implications

So with cyberbucks an information provider is free to split her information into the finest hairs imaginable. She can sell it by the article, the page, or the paragraph—all without worrying about the resulting "micropayments." Moreover, the cost of providing information is driven down. That removes the barrier to entry for many information providers and increases the number of players in the market.

That's the good news. The bad news is that the buyer is now disseminating more and more information about himself throughout the system. Since there are so many more transactions being performed, there are also so many more points at which information about a buyer can be garnered. Before, a buyer may have purchased a single subscription to an information source. Information about himself remained with a single source. With digital cash, the buyer is spreading that information around Silicon City like a graffiti artist. People complain now that their names are being sold by magazines and mail-order houses to anyone with the money to buy a mailing list. Imagine the hue and cry when that activity is increased a gigafold. Privacy issues are bound to come to table.

One of the biggest concerns here is that something large institutions call e-cash will be passed off as real digital money. This pseudo digital cash would be one step removed from a credit card purchase and a prelude to banning cash altogether. Control agents, like governments, banks, and corporations, wouldn't mourn the death of hard currency if what replaced it was something that gave them the power to trace all purchases without a buyer's or seller's permission or knowledge, report spending patterns automatically, target lists of those who frequent outlawed businesses, and surveil all interpersonal economic transactions.

"There was a time when I would've said Americans, at least, would've rejected such a thing," wrote one Net surfer. "Too many memories of 'Papieren, bitte. Macht schnell!' But I now think most Americans (and Europeans) are so used to producing documents for every transaction, and so used to using Visa cards and ATM cards at gas stations, supermarkets, and even at flea markets, that they'll willingly—even eagerly—adopt such a system."

Live Free in an Unfree World

http://www.valleynet.net/~wealth/tlhome.html

Live Free in an Unfree World
TERRA LIBRA
The World's First Truly *Free* Country!

- Project Terra Libra
- Articles, Essays, Reports
- Terra Libra Chamber of Commerce
- Personal Power Institutes
- What People Say About Terra Libra
- The Terra Libra Catalog

13 THE FUTURE

Science fiction used to be something far removed from daily experience. But with computers appearing everywhere, rhizomes from the Net touching millions of households, and technology morphing faster than a face in a shaving advert, today's SF could be the next day's experience. The reality of a shifting reality has many people squirming in their desk chairs. Among those squirming most anxiously are the people assigned the task of describing reality to us on a daily or weekly basis: the folks in the media. But even the media sometimes have to turn to speculative fiction to get ahead of the curve on a trend like the development of cyberbucks. That's what *Time* magazine did when it published a story by Neal Stephenson, author of *Snow Crash* and *The Diamond Age*, called "The Great Simoleon Caper" in the Spring, 1995, issue. (See reprint at the beginning of this book.)

Currency as Investments

Neal Stephenson's story may be speculative fiction with a wry twist, but it isn't far removed from what is lying in our path as digital cash begins to enter the financial freeway of nations.

Jon Matonis, whose free-market views we outlined earlier in these pages, doesn't see national currencies disappearing as private e-cash flows into the global economy. Here Matonis speculates on what a simple transaction would be like in the year 2005:

"The year is 2005. I buy lunch at a deli and I pay in wireless digital cash from my electronic wallet....The cashier gives me a choice of monetary units which are both displayed on the flat-panel screen for me to view. My turkey-and-cheese sandwich will cost me US $50 or 5 pvu. The monetary symbol 'pvu' is an abbreviation for 'private value units,' which now compete in most commercial settings with the US dollar and have stayed remarkably stable since their initial issuance in mid-1996."

As Stephenson sagely points out in his story, the acceptance of digital cash depends on people accepting what's behind the money, the "assets." Consumers will begin to look at currency as just another investment in their personal portfolio. This will probably turn into a selling point for minters as they vie to attract consumers to their currency.

Matonis observes: "Suggestions for monetary backing include equity mutual funds, commodity funds, precious metals, real estate, universal merchandise and/or services, and even other units of digital cash. Anything and everything can be monetized. This will undoubtedly develop into a main basis for competition among digital cash providers as each one promotes their underlying currency backing as the strongest and most reliable. Unlike today's national monetary systems, the benefits of a strong currency will be immediately noticeable within a country's borders. With multiple monetary unit providers, domestic prices will adjust rapidly to reflect relative values of monetary units and the holders of stronger currencies will benefit. This is a vastly different world than we have now and consumers will analyze currencies as the investments that they really are."

This sounds like equity management players like mutual fund companies might have an upper hand in the private e-cash systems of the future. But that isn't the case, according to Matonis. "Mutual funds of mutual funds exist today and contracts can be executed with the specialist managers of those funds," he writes. "Outsourcing the portfolio function takes advantage of the experts in the field today who compete already on reliability and overall performance—prime benchmarks for a private monetary unit. The issuer's skills should concentrate on distribution, monitoring geographic circulation of the unit, and managing the rate of redemption."

Clearinghouse for Cleaning House

Managers of private digital cash systems will be under even more pressure to play with the value of their currency than their public counterparts because they'll be operating in an ultracompetitive marketplace. What will probably develop to moderate this activity are "clearinghouse parties" who will perform the functions that central banks perform in the public systems; that is, make sure issuers of digital cash maintain an adequate balance between e-cash outstanding and the reserves behind it.

"Systems of clearing and redemption are a necessity for the smooth operation of free banking as they provide a check on overissuance and the general deterioration in sound credit," Matonis explains. "Therefore, the manager of a private monetary unit can rely on these clearinghouse parties to communicate to the public the unit's standing in the economy. Moreover, if the discount of a particular unit begins to deteriorate, it can alert management to the fact that some market forces are affecting the demand for that unit."

Writer's Market

One wonders how much money Neal Stephenson might have made on his "Simoleon" story if he posted it on the Internet and could collect 20 cents a hit for it. Given the author's following, 50,000 hits would probably be a low estimate. Right now, those thoughts are idle speculation. But when digital cash arrives on the Net and microtransactions become a reality, those numbers will be dancing in many a writer's head.

According to the National Writers Union, "income-producing self-publishing on the Net could be a great boon not only for freelance writers, but for readers as well. The ability to earn a living from online distribution of one's work will encourage a wider range of writers to produce a wider range of materials for a wider range of audiences. This objective is especially important at a time when commercial print publishing continues to grow more concentrated, and corporate publishers are seeking to establish a dominant presence in electronic publishing as well."

For writers, e-cash may be more than a nice perk courtesy of the Information Age; it may be a prescription for their future mental and financial health as well as the survival of the independent scribbler. "By encouraging online commercial self-publishing," the union contends, "we hope to prevent the complete corporate domination of the Web. The ability of individual writers to make their work available on the commercial part of the Net will create a much-needed diversity of voices. It could also significantly reduce the financial frustration of being a freelance writer."

If a writer is going to sell his work online, he's going to have to get the word out to other Web crawlers. Technically sophisticated writers may try to go it alone and set up their own Web sites. Other writers may form consortiums or partnerships where writers of similar or related interests could band together at a site to increase their draw. Yet another approach will be transaction services, which would be acting as a kind of online newsstand, bookstore, or distributor.

Regardless of how a writer decides to circulate her material, one thing will be certain: The micropayments she receives for her work will go a long way toward settling those arguments in the newsroom about how much the "talent" is worth, or how widely read an author is.

Volume will be important to the online author. As the writers' union points out, online sale of information and other textual material will not succeed unless purchase prices are kept very low. Transaction systems should be developed that can accommodate tiny payments, even as low as a few cents for small quantities of material. Keeping prices low will discourage unauthorized reproduction and make it easy for buyers to respect the copyright of the sellers.

Privacy will be as important for the online publisher as it is for the digital cash transaction itself. Readers should have the ability to purchase electronic materials in a way that is not traceable. Just as purchasers of certain materials from bookstores and newsstands may choose to use cash and thus leave no record of their transaction, so too should that option be available for online commerce. After all, what reveals more about a person than what he reads?

"The prospect of online self-publishing, in particular, represents a dramatic development for writers," the union states. "The absence of significant start-up costs and printing expenses, along with the possibility of easy revenue collection, means that independent publishing would be feasible for a much larger number of writers. For the first time since the days of Ben Franklin, writers—as opposed to publishers, printers, and distributors—would receive the bulk of the money that readers pay for their work, and they would get that money promptly. For a group of people used to being paid too little too late, online commerce could be a godsend."

If anyone finds out if electronic commerce will be a godsend to writers, it will be Newshare Corporation (http://www.newshare.com), a venturesome firm in Williamstown, Massachusetts. Newshare is developing a software package called Clickshare aimed at publishers who need to collect micropayments for their online information.

Clickshare resides on a Web server. It clears transactions much the way the credit card clearing system clears transactions. When a buyer purchases information through Clickshare, the program takes care of the accounting. The purchase price of a piece of information can be divvied up any way the owner of the server desires.

Moreover, Clickshare works across servers so a buyer can purchase information from any Clickshare-enabled server, and it will settle up with all the servers involved. So if you log on to Clickshare through server A and buy information on server B, the program will charge server A for the transaction. At the end of a billing period, the owner of server A can pay the owner of server B for all the purchases server A's Clickshare users made on server B.

For example, a subscriber to an online newspaper in San Jose, California, might wish to download a music review by a writer who has posted the review on a server in Massachusetts. If both servers are Clickshare-enabled, the subscriber can purchase the review without having to register at the author's server.

After Newshare takes its cut—15 percent of the transaction—the author can receive payment for the transaction and the San Jose newspaper can get a commission for facilitating the transaction.

Both the newspaper and the author receive monthly reports on how many people have accessed the review. They also receive monthly payments—royalties for the writer, commissions for the newspaper. Newshare receives a taste for its role in facilitating the transaction, the same way a credit card company takes a piece of the action at the retail level. For example, a 10-cent purchase may generate 3.5 cents for a subscriber's "home" Clickshare server, 1.5 cents for Clickshare, and 5 cents for an author.

Initially, charges for these services will probably be cleared through the credit card system. For example, if you subscribe to the San Jose server for $5 a month, the $5 is charged to your credit card. If you have

> **For the first time since the days of Ben Franklin, writers—as opposed to publishers, printers, and distributors—would receive the bulk of the money that readers pay for their work.**

Clickshare charges, they could be added to that monthly charge. So if you made twenty Clickshare transactions worth $2, your credit card account would be charged $7 for that month instead of $5. "By aggregating all those charges to a single charge to a credit card, the 25-cent transaction fee and the 2 percent discount rate that's typically charged for a credit card transaction are spread out over many transactions instead of a few," explains Bill Densmore, president and cofounder of Newshare. "For that reason, it's possible to charge a discrete amount of pennies for information access."

According to Newshare, the system is economical for purchases as small as 10 cents.

Clickshare is designed to be browser-independent. That means that whoever's Web browser you use, you will be able to use Clickshare-enabled servers. However, initially the software will only be available for Unix-based servers.

Servers using Clickshare communicate with a server at Newshare, where the Clickshare accounting is handled. Also stored on this server will be user profiles. The information in these profiles can be customized by the user. He could block out porno sites, for example, or refuse to look at advertising.

Densmore says that Clickshare isn't designed to compete with other payment systems like First Virtual and CyberCash. "We are a niche implementation," he notes. "Our niche is transactions that are too small to be implemented by payment systems that clear every transaction through the credit card system. A user could use First Virtual or CyberCash to pay their Clickshare bill."

Not only does Clickshare clear transactions much as the credit card system does, its security has a credit card flavor, too. When a user logs on to a Clickshare session, she's assigned a unique number for that session by the Clickshare server. The number is sent from the Clickshare server to the user's server in encrypted form. Clickshare sessions also have a time limit. So if that session number is hacked, the most damage an information highwayman could do was run up charges for one session's time limit, which is akin to the $50 ceiling on unauthorized credit card charges. "We don't think that kind of hacking will occur routinely," Densmore says. "And even if someone figured out a way to do it twenty-four hours a day, the shrinkage would be in the same order of magnitude that goes on every day in the cellular phone industry. That hasn't kept the cellular phone industry from being profitable and growing very nicely."

Densmore sees Clickshare initially appealing to niche content providers more than large information organizations such as newspapers. "Any one newspaper wouldn't benefit much from this," he says. "It would take the involvement of a large wad of newspapers for it to work really well for them. It's more likely that niche providers will gather together in similar niches and start using it initially."

Online services are a possible outlet for Clickshare, but they would have to change their present mind-set to accept the idea, according to Densmore. "They would have to accept the idea of letting their users escape from the walls of the online service," he says. "And they would have to let people enter their services and buy content à la carte.

"But the online services you see today won't be anything like this a year from now," he says. "They will be largely super Internet Service Providers. Otherwise, they're going to be like phone companies that are not part of the phone grid. Who wants to be part of the Tinkerbell phone network if Tinkerbell doesn't connect anywhere else in the world? That's where they're going to be a year from now. Even the Microsoft Network is going to be in that position."

A Collection of Kangaroo Courts

In Major League Baseball, some clubs have a tradition called the kangaroo court. A judge is appointed at the beginning of the season. From time to time, the court is convened in the clubhouse before a game and fines are meted out for pseudo-crimes and goofy misdemeanors. A similar judicial system may arise when e-cash becomes widespread.

As the Internet grows, it may be necessary to set up a system of fines to enforce netiquette. Old timers on the Net may lament the loss of the "good old days" when regulatory systems and fines were not needed to enforce netiquette, but human nature being what it is, when the norms and values of a society are not reinforced by legal and economic incentives, the norms and values tend to break down over time. Some newsgroups may feel that they do not need pricing or regulations to insulate themselves from violations, but in other cases the outlaws will overrun the town if something isn't done.

Arnold King, in an online article written for GNN (http://nearnet.gnn.com/), described such a regulatory system. Newsgroups could form regulated newsgroup associations. To post to the newsgroup, a Net surfer would have to agree to abide by the charter of the group. That includes paying e-cash fines for violating the charter. The regulated newsgroup could do anything it wanted with the money from the fines, such as donating it to libraries in low-income neighborhoods to provide Net access to have-nots.

Each newsgroup would set its own schedule of fines and method of enforcement (perhaps a rotating three-member board could rule on any post). While a minor infraction against a newsgroup's charter might cost only 25 cents, a major violation could cost $10. This might deter spam artists such as Canter-Siegel, the law firm that bombarded 6,000 newsgroups with ads, from flooding newsgroups with irrelevant postings. The vast majority of people, who post legitimate messages, would continue to post for free, without any hassles.

Similar tactics could be used to combat Internet junk mail. If you send someone unsolicited e-mail, the e-mail will tell her how much you will pay her to read it. If you will not pay her, then she may or may not read it. If you will pay her, say, $2 to read it (she will signify that she is doing so by checking certain boxes in your message as she goes through it), then she will have a look.

A Day in the Life

What will a day be like for an average citizen in e-cash society? Let's follow the activities of the Chips family one Sunday in the future.

The smell of bacon from the downstairs kitchen wakes Mr. Chips. He silently thanks Mrs. Chips for letting him get a few extra winks before he has to leave his cozy bed. The bacon reminds him of something else: pigskins. He forgot to play his football card yesterday. He hits a preprogrammed button on his bedside phone and connects to an offshore betting parlor. He pulls a smart card from his wallet, slips it into a slot in the phone, and punches numbers in response to menus appearing on the phone's liquid crystal display. When he's through with the menus, he uploads his $2 bet from his smart card to the betting parlor.

One advantage of being the last one out of bed is that there's no waiting for the bathroom. Mr. Chips is looking forward to a hot shower to get his sluggish blood moving on this sunny morning. He slips his smart card into a slot by the shower stall and begins running the water. With global warming and frequent water shortages, the town water department has installed real-time meters in every house. Everyone pays for their water as they use it. When Mr. Chips finishes his shower, the cost of the water is deducted from his smart card and the e-cash immediately transferred via wireless link to the water department's account in a local bank. Mr. Chips slips his smart card into his card reader. His shower cost 2.371 cents.

Groomed and dressed, Mr. Chips descends the stairs, ready for breakfast. At the breakfast table, the kids, Kayla and Kyle, have already started eating without him. He sits at the table, breakfast at the ready, news slate by his plate. He turns on his slate and checks out the daily index. He finds the stories he wants—last night's sports action, some business analysis pieces—and requests a download. He's instructed to slide his smart card into the slate. The stories are downloaded and they're paid for with e-cash from the card. A message appears on the slate informing him of the total cost of the transaction, 30.125 cents. The money is immediately distributed into the accounts of the involved parties: the publishers, distributors, and writers. Mrs. Chips says something to Mr. Chips, but he doesn't respond. He's too engrossed in his news slate and cholesterol-free eggs from genetically engineered chickens.

After breakfast, the family piles into the car to go to church. To start the car, he once again uses his smart card. The card activates a computer in the car that keeps track of the mileage. The car is leased and the family pays for it on a per-use basis. Each time the family uses the car, the cost of the trip is deducted from the e-cash total on the card.

When the family arrives at church, Mr. Chips decides to park at the church lot. It would be cheaper to park on the street and pay for metered parking with e-cash, but the church gets a cut of the parking proceeds and it's another way to help out the parish. At the entrance to the church parking lot, Mr. Chips slides his smart card into the silent attendant. The time he enters the lot is recorded on the card. When he leaves the lot, his time there will automatically be calculated and the cost of parking downloaded in e-cash from the card.

Before leaving the car, Mr. Chips takes out his digital purse and asks the kids for their smart cards. He loads them with cash so they can make a donation to the church when the "basket" is passed through the congregation during the service. Although the church still uses a basket for tradition's sake, inside the basket is an electronic purse that allows church members to make their donations in cyberbucks. Digital cash payments are usually anonymous, but most members of the congregation remove that feature from the money they give to the church. They want the minister to know who gave what to the church, and they want a public record for tax purposes. As Mr. Chips loads up the kids' cards, he makes sure to restrict the spending criteria for the cash so it can only be used as a donation to the church.

After church, the kids say they want to go to the movies. Mr. Chips downloads more cash into their cards, making sure to program it only for movie theater expenses. In addition to restricting individual downloads, the kids' cards have other built-in restrictions. The card won't dispense e-cash for a number of things their parents don't want them to buy such as cigarettes, booze, and time on the adult cable TV channels.

The kids are dropped off at the movie house and Mr. and Mrs. Chips head home for a laid-back Sunday afternoon. Mr. Chips plans to sit in front of the media center and watch pay-per-view football, paid for with, what else, e-cash. Mrs. Chips plans to spend the afternoon reading on the deck. She's downloaded and paid for in digital cash a new book by an old-time science fiction author named Neil Stephenson.

14 GETTING INVOLVED

We have seen what digital cash is, who some of the players in its development are, and what its potential is. But what if you want to get a piece of the action? This chapter offers some detailed information about how to throw your lot in with some of the companies on the leading edge of digital cash development.

First Virtual

First Virtual Holdings has some of the most extensive documentation available on the Internet about how to buy and sell within its framework.

Getting Started As A Seller

Whatever it is you're thinking of selling—reports, essays, stories, columns, poems, software, artwork, photographs, research results—if it can be produced in electronic form and distributed over the Internet, you can use the First Virtual Internet Payment System to sell it.

Before you start selling through First Virtual, you need the following:

- A private e-mail account.

- A checking account in a United States bank, into which First Virtual can deposit the proceeds of your sales.

The First Virtual account application process is simple and straightforward:

You give First Virtual basic information about yourself, such as your name and address, over the Internet. You can do this on the World Wide Web right now, or send e-mail to apply@card.com to request an application in e-mail.

You activate your account for selling by sending a check for $10 through the mail to the First Virtual address provided to you in your e-mail confirmation message. The information on the check is used to identify the checking account to which you wish the proceeds of your sales deposited, and the $10 pays part of the cost of setting up your account.

You wait for the e-mail message from First Virtual confirming that your account has been activated for selling and providing you with your account identifier, which you can use to build a storefront on the First Virtual InfoHaus or to submit transactions to First Virtual for settlement if you make your own sales.

Here are some specific matters you should be aware of when setting up your account for selling:

You may open as many First Virtual accounts as you wish.

Any First Virtual account may be activated for selling, for buying, or for both. If you plan to open a storefront on the First Virtual InfoHaus, you must activate your account for buying (by following the simple instructions in First Virtual's e-mail confirmation message, to link it to a credit card account) as well as for selling.

When you apply for your account, you must specify your "full name." The name you provide will be visible to people who buy things from you. You are not required to use your actual name; you may use a pseudonym or a business name. If you will be setting up an InfoHaus storefront, First Virtual encourages you to use the same name for your storefront that you use when you apply for your account.

If you want to get going without waiting for your check to be received and processed by First Virtual, you may begin using your account for selling, or use it to open an InfoHaus storefront, as soon as you have

Getting Involved

activated it for buying—a process which usually takes less than a day to be completed. The proceeds of any sales you might make will be automatically held in the account until it is activated for selling.

Once you have activated your account, you can begin accepting First Virtual for payment immediately. "Accepting First Virtual for payment" means:

- You clearly post the prices of your information products on your server, or clearly state the prices when discussing the products with potential buyers.

- When a buyer requests a copy of one of your information products, you require him or her to provide a First Virtual account identifier.

- If you wish, you verify the validity of the First Virtual account.

- You deliver the requested information product to the buyer.

- You submit the required information about the transaction to First Virtual to initiate the collections and payment process.

- The buyer confirms the purchase with First Virtual by e-mail, and you are sent an e-mail notification about the confirmation.

The InfoHaus was deliberately designed to allow buyers to review information before deciding whether to pay for it. Consequently, for sales using the InfoHaus, the merchandise must be delivered before the buyer has confirmed the purchase.

A very few First Virtual merchants have chosen to initiate the collections and payment process first, and wait for the buyer to confirm the transaction with First Virtual by e-mail—i.e., to commit to pay—before delivering the requested product. First Virtual strongly discourages sellers from doing this, and requires them to make such a policy clear to their customers if they decide to adopt it. You may submit a transaction manually to First Virtual, using e-mail or Telnet. You will need to provide the following information about each sale:

- The buyer's First Virtual account identifier.

- The seller's First Virtual account identifier (i.e., yours).

- The amount of the sale.

- The currency (e.g., U.S. dollars).

- A description of the transaction (e.g., the name of the item purchased).

First Virtual can also provide you with software you can install at your own site, which modifies your existing WWW or FTP server to automate the sale of information and the submission of transactions using First Virtual.

Finally, the sale of information using First Virtual can be automated using scripts you write yourself. For example, if you already operate a mail archive server which sends out files in response to messages, you can modify it easily to charge for files by requiring that a request for a file include a First Virtual account identifier to be charged.

The InfoHaus is First Virtual's public-access information server, which operates a World Wide Web site (at http://www.infohaus.com), an FTP site, and an e-mail information server.

Anyone with a First Virtual account can set up a storefront on the InfoHaus easily using common tools like e-mail and Telnet. If you use an InfoHaus storefront to sell your information products, the InfoHaus takes care of billing and collections for you automatically. The free *First Virtual InfoHaus Seller's Guide* is a complete step-by-step guide to setting up an InfoHaus storefront. You can request more information about the *InfoHaus Seller's Guide* by sending an e-mail message to infohaus-guide@fv.com, or can view it online.

The First Virtual Internet Payment System is a transaction tracking, collections, and settlement system. A seller submits transaction information to First Virtual; on the basis of that information, the appropriate buyer is contacted by e-mail and asked to confirm the payment; once confirmation is received, the buyer is charged and the seller is credited.

Each time a buyer downloads one of your products from the InfoHaus—by retrieving the product's "paid" portion from your InfoHaus Web page, retrieving it via FTP, or purchasing it via e-mail—a transaction is automatically initiated by the InfoHaus.

Once the InfoHaus initiates the transaction, First Virtual contacts the buyer by e-mail. First Virtual's e-mail message to the buyer—which is not copied to the seller—gives the buyer the "full name" from your First Virtual seller's account, the price of the item, and a description of the item, and asks the buyer to confirm the purchase. If the buyer does not reply, First Virtual repeats the message within a few days. (If the buyer

Getting Involved

does not reply to repeated inquiries, first his or her account is suspended, and then, after about a month, he or she is deemed to have declined the purchase.) The purchased item is referred to by its one-line description, if you submitted one when you installed the item in the InfoHaus; if you didn't, it's referred to by the "info name" that identifies it in the InfoHaus—for example, "photo.nature.lions."

After the buyer replies, a confirmation message called a "transfer-result" is generated, which confirms that the buyer has agreed to pay or indicates that the buyer has declined to pay. This e-mail message, which is copied to you, is your indication that a sale has been made.

Payment will be made by direct deposit to the checking account with which you activated your account for selling. The frequency of payment depends on the volume and amount of your transactions.

Payments are "aged" by First Virtual for a reasonable amount of time to enable the company to be paid by buyers in order to be able to guarantee payment to sellers. Before First Virtual deposits the proceeds of your sales into your checking account, a standard First Virtual transaction fee of 29 cents, plus 2 percent of the selling price, is deducted from the sale price of each product sold.

Some of the InfoHaus fees discussed below are being waived for any sales made while the InfoHaus is considered to be in its beta test phase. The "Message of the Day" on the main InfoHaus page at http://www.infohaus.com always clearly states which fees are in effect.

If you choose to make your sales on the InfoHaus rather than from your own server, an additional InfoHaus handling fee of 8 percent of the selling price is deducted from the proceeds of each sale made in this way. This means that if you use the InfoHaus for your sales, you will be credited as in the following examples:

- If your product costs $1, a fee of 29 cents plus 2 cents (2 percent) plus 8 cents (8 percent) will be deducted, and 61 cents will be paid to you.

- If your product costs $10, a fee of 29 cents plus 20 cents (2 percent) plus 80 cents (8 percent) will be deducted, and $8.71 will be paid to you.

- If your product costs $100, a fee of 29 cents plus $2 (2 percent) plus $8 (8 percent) will be deducted, and $89.71 will be paid to you.

- A payment processing fee of $1 is deducted each time a deposit of proceeds is made to your checking account; that fee, which reimburses

> **Although First Virtual does not require that you let buyers download your information before making a commitment to pay, it feels strongly that it is in your interest, as well as the buyer's, to do so.**

First Virtual for the cost of using the Federal Reserve's Automated Clearing House (ACH) system to make your deposit, is per deposit, not per transaction; it remains the same regardless of the amount of your deposit or the number of transactions it represents.

Finally, if you use the InfoHaus, you are charged a nominal storage fee of $1.50 per megabyte (or portion) per month for the disk space you use. This fee is the only one that will be charged each month whether or not you make any sales; in the event that you do not make enough sales to cover your storage fee, the remainder will be charged to your First Virtual account for your review. (Declining this charge will disable the use of that shop on the InfoHaus.)

In order to make sales from your own server, you need to collect the buyer's First Virtual account identifier and then initiate each transaction yourself by giving First Virtual certain pieces of information about the transaction so that the confirmation and settlement process can begin. This process can easily be automated. First Virtual provides software you can install on your own server, or you can write your own programs or scripts.

Although First Virtual does not require that you let buyers download your information before making a commitment to pay, it feels strongly that it is in your interest, as well as the buyer's, to do so, and it strongly encourages you not to require payment in advance. Partly, this is because requiring payment in advance goes against the Internet community's strong commitment to the open exchange of information. But there are practical considerations as well. You're free to set any terms of sale that you and your buyers can agree upon, but First Virtual thinks that the terms it suggests, and that have been chosen for the First Virtual InfoHaus—namely, letting buyers examine your information before having to decide whether to pay for it—will result in substantially more sales and higher revenues for you.

If you require payment in advance, a very large proportion of your prospective buyers will simply disappear, and your revenues will drop accordingly. Most people will be unwilling to commit in advance to paying for something they haven't seen. You should be reassured by the fact that under the economics of Internet commerce you lose little or nothing if a few buyers examine your information and then decide not to pay, because it costs you nothing to "manufacture" another copy for the next buyer, and you don't have to pay for returned merchandise, repairs, or restocking, as you would if you were running a store in the real world.

Furthermore, First Virtual's tight controls on abuse ensure that any buyer who takes advantage of the goodwill of its sellers will have his or her First Virtual account suspended or terminated.

You may, however, choose to wait to deliver information to buyers until after they have received and replied yes to First Virtual's confirmation message. You also may wish to verify each buyer's First Virtual account to ensure that it is valid before delivering information to the buyer, or before submitting a transaction for settlement. There are several simple ways you can do this:

- Finger: If you have access to the common information utility known as Finger, you can use it to verify that a First Virtual account identifier is valid by fingering the account on card.com. You should leave out any spaces, commas, and other nonalphanumeric characters in the account identifier when you type it. For example, if a buyer claims to have the account identifier "Gold Finger #8," you can verify that it is valid by typing "finger goldfinger8@card.com" (or the equivalent command on your machine).

- Telnet: If you have access to Telnet, you can use it to connect to telnet.card.com and view First Virtual's menu. Select the appropriate menu choice to request an account inquiry and enter the account identifier to learn the status of the account.

- E-mail: You can send an e-mail message to inquiry@card.com, putting the account identifier you wish to verify into the subject line of your message. First Virtual's server will reply to you via e-mail with status information for the account.

- Using the FV-API: If you have the "fv" program installed (available from ftp.fv.com under pub/code/fv-api), you can use fv check@card.com account to see the status of an account.

There are several ways you can initiate a transaction; in each case, you will need to provide the following information:

- The buyer's First Virtual account identifier.

- The seller's First Virtual account identifier (i.e., yours).

- The amount of the sale or the currency (e.g., U.S. dollars, abbreviated "USD").

- A description of the transaction (e.g., the name of the item purchased), which can be up to forty characters.

You can automate the process of initiating a transaction in several ways:

- You can write a program or script that composes properly formatted "transfer-request" e-mail messages and sends them to First Virtual.

- You can use software provided by First Virtual, or software that you write yourself to conform to the Simple Green Commerce Protocol (SCGP) specification, to submit transactions via a direct connection.

- You can design HTML forms that automatically send their contents to a special First Virtual address for processing.

You can submit a transaction via e-mail by sending a message to "transfer@card.com." This message, the "transfer-request," must have at least the five items noted above, formatted as in the example below, in the message body.

To: transfer@card.com
From: (you)
Subject: (anything you like)
BUYER: flap-goldfinger #8
SELLER: blue-Tormented Intellectual
AMOUNT: 5.00
CURRENCY: USD
DESCRIPTION: "The Ravioli Chronicles"—short stories. Thank you for buying "The Ravioli Chronicles"! If you liked it, you'll love the sequel, "Spaghetti Stories," available from Pasta Press for $7.50—send e-mail to pasta@somewhere.com for information.

Please note the following items:

- Each field's name (such as BUYER) must begin at the left margin, and be followed by a colon and a space.

- You may omit punctuation and spaces from the First Virtual account identifiers, but you need not do so.

- You should not include a dollar sign or any other currency symbol in the AMOUNT field.

- In the CURRENCY field, United States dollars is abbreviated "USD."

Getting Involved

- You have up to forty characters for the description field.

- There must be a blank line before your shop description, as in the example. If you put additional text into the body of a transfer-request, as in the example above, it will be delivered to the user as a "message from the seller." You must insert a blank line before your message, as in this example. In your message, you might put a longer description of what you sold, or a plea to donate to a worthy cause, or a description of other products available from you, or whatever else you'd like.

There are also a number of optional fields you can supply:

- One is TRANSFER-TYPE:, which should be "info-sale," "cost-recovery," "donation," or "usage-fee." This field is currently ignored by First Virtual, but if you use it, it's passed along to the buyer in the confirmation message from First Virtual.

- Another is TRANSFER-ID:, which is a short string between angle brackets that you can use to identify the transaction for your own records; you may use letters and numbers, periods, and hyphens, and it must have one @-sign in it, like this: <bill.09.12.1994@mydomain.com>.

- A third optional field is DELIVERY-STATUS:, which should be either "pending" or "delivered." This, again, is ignored by First Virtual but is passed along to the buyer if you use it; it defaults to "delivered."

You can also initiate a transaction by connecting via Telnet to fv.com and selecting the menu command for a "funds transfer." Using the Telnet interface, it is not possible to include the optional fields noted above, nor is it possible to include a "message from the seller."

Making use of software and specifications provided by First Virtual, you can write scripts and programs that submit transactions automatically.

Assuming that the transfer-request you submitted was properly formatted and contained all the required information (if not, you'll soon receive an error message by e-mail), First Virtual contacts the buyer by e-mail.

First Virtual's e-mail message to the buyer—which is not copied to the seller—gives the buyer your "Full Name" as you submitted it when you applied for your First Virtual account, and the amount and description you supplied, and asks the buyer to confirm the purchase. Any "message

Soon First Virtual will begin to offer to debit buyers' checking accounts directly, but right now that technology is not in place.

from the seller" that you provided is also included.

If the buyer does not reply, First Virtual repeats the message after a few days. (If the buyer does not reply to repeated messages, his or her First Virtual account will be suspended; eventually, after about a month, he or she will be considered to have declined the transaction.)

After the buyer replies, the e-mail confirmation message, or "transfer-result," is generated. It is copied to you to let you know whether the buyer has agreed or declined to pay. (If you are automating processing of transfer-results, you may wish to use the contents of the "Authorization" field, which will be "yes" if the buyer confirms, "no" if the buyer declines, "timeout"—equivalent to "no"—if the buyer fails to answer, or "fraud" if the buyer replied that the transaction was unrecognized.)

Getting Started as a Buyer

There are a few restrictions on who can sign up for a First Virtual account and use it to buy information.

- First, you need a personal e-mail account with an address and password that are private to you alone.

- Second, for now, you need a valid credit card. Soon First Virtual will begin to offer to debit buyers' checking accounts directly, but right now that technology is not in place. First Virtual currently handles Visa and MasterCard. Your credit card number is never sent over the Internet, and is never stored on a machine that is connected to the Internet. Also, except for the onetime account setup fee of $2, and a nominal fee when you request First Virtual to change or update your credit card information, a charge is never posted on your credit card without explicit confirmation from you via e-mail that you agree to it.

- Third, for now, you must use a credit card which is able to make charges in United States dollars. Soon, however, anyone in the world will be able to buy and sell information using First Virtual, in their own currencies.

When you apply for your First Virtual account, you'll be asked for three important pieces of information.

- First, you'll be asked for your "Full Name." This is the name that people will see who buy information from you, or sell information to you. You may use a pseudonym.

- Second, you'll be asked for your electronic mail address. As explained above, First Virtual communicates with you through your electronic mailbox, and this mailbox should belong to you and only you.

- Third, you'll be asked for your "ID-Choice"—the string that your First Virtual account identifier will be based on. Your account identifier is the string that will identify you to sellers and to First Virtual when you buy information. First Virtual, unlike many other Internet services, allows you to choose most of your own account identifier. Since you will be giving this account identifier out over the Internet as part of every First Virtual transaction, you shouldn't use any secret or sensitive information (such as your e-mail password, your date of birth, etc.) that you wouldn't want to be giving out to random strangers. And since you want to keep it relatively confidential, you shouldn't use anything that's too easy for people who don't know you to guess (like your name or your e-mail address). For your own protection, First Virtual will change your ID-Choice a little bit in generating your actual account identifier.

When the application process is complete, a confirmation will be returned via e-mail with your actual account identifier. Your account identifier will match the ID-Choice you specified, except that it will have a short random word added to the front to assure its uniqueness and to make sure it's difficult to guess. Your account identifier is safe to send over the Internet because it is linked to your credit card number only deep within First Virtual's system, on a machine that is never connected to the Internet and is managed by its financial transaction partner. It also reveals nothing about you. No seller (or other person) can use your account identifier to discover your postal address, your phone number, or even your e-mail address (although a seller may be aware of your e-mail address if you connected to his or her information server). Sellers are told your full name, but only for convenience and security, so that First Virtual can refer to you without sending your account identifier over the Internet more often than necessary. And if your account identifier were ever stolen, you would still be protected, since First Virtual asks via e-mail for your explicit permission to make any charge to your credit card. If you don't recognize the item that First Virtual is asking to charge you for, the company will invalidate your account identifier and let you apply for a new one. So the worst that anyone can do who knows your account identifier is to make you go through the annoyance of having to change it.

There are a number of ways to give First Virtual your name, e-mail address, and ID-Choice.

You can send an e-mail message to apply@card.com, and First Virtual's e-mail server will send you a simple application form automatically. When you get the application form, simply save the message in a file and follow the instructions provided in the message to fill in the blanks. Then send the edited message back to First Virtual at newacct@card.com. As soon as the message arrives, First Virtual will start processing your application.

Or, if you prefer, you can use Telnet to connect to telnet.card.com to fill out the application interactively. (If you happen to be running X-Windows, you will be able to use a graphical interface to fill in the form, but neither X-Windows nor Unix is required; any ordinary Telnet connection is fine.)

Second, give First Virtual your financial information. For your own protection, First Virtual will not ask for your credit card number or other sensitive information over the Internet. Instead, once First Virtual has started processing your application, you will receive an e-mail message from it telling you to call its computer on the phone and give it your credit card number using your Touch-Tone keypad. You can make this call anytime, day or night. The e-mail message will give you two things: a toll-free 800 telephone number to call (with a direct-dial number in case you are calling from outside the United States and Canada), and an "application number." This application number is just a randomly generated number that you use to tell First Virtual who is calling. Have your application number and your credit card ready when you call. A computer will ask you for each number in turn. Once the numbers are successfully entered, you will receive another confirmation via e-mail, as described below.

In a few hours, once the confidential information is moved to the proper machines, you will receive one more e-mail message from First Virtual. This will acknowledge that your account is ready to use. It will also reveal the random word First Virtual put at the front of your ID-Choice to form your account identifier. With this information in hand, you are all ready to go out on the Internet and buy information using First Virtual.

Making an information purchase is easy. As usual on the Internet, finding the information you want can be harder than getting it once you've found it.

Every day, new merchants are setting up shop on the First Virtual InfoHaus, and companies are adding First Virtual payment capabilities to their own servers. Since the First Virtual system is public, it does not know about new sellers using the system unless they tell First Virtual.

Information for sale on the First Virtual InfoHaus and most other servers will always be presented with a description (provided by the seller) and a price (set by the seller). The seller's price is the price you pay; First Virtual never charges you any fee for a purchase transaction.

In keeping with the Internet's history of open information exchange, most sellers will allow you to download and review their information before requiring you to decide whether to keep and pay for it. (A few sellers may require you to agree to pay for information before they will let you download it, but that's something First Virtual discourages.) When you find a seller offering something that looks interesting, at a price that looks reasonable, tell the seller's information server that you'd like a copy. Usually, the seller's server (or the InfoHaus) will ask you to enter your account identifier before allowing you to download information; this is done so that the transaction can be reported to First Virtual, who can then ask you whether you agree to pay for the information, as described below.

Usually, when you log in to a First Virtual–compatible FTP server, you should use "fvftp" as your user name rather than "anonymous." Then give your First Virtual account identifier (rather than your e-mail address) as your password. You can generally browse directory listings—which will tell you how much the files cost—for free. (If there is ever a charge for browsing, the server will notify you.) The prices of any files you decide to download will be billed to your First Virtual account, subject to your approval via e-mail.

Sometimes when you attempt to follow a link (e.g., by clicking on a word), instead of the information you expect, you will see a screen telling you the price of the information and requesting your First Virtual account identifier. If you provide your account identifier, you will have access to the information, and your First Virtual account will be billed for its cost, subject to your approval via e-mail.

Some Internet mailing lists may charge a subscription fee that is payable through your First Virtual account. If you ask to subscribe to such a list, you will receive a message telling you the subscription fee and asking for your First Virtual account identifier. If you provide your account identifier, you will be added to the list and your account will be billed for the subscription fee, subject to your approval via e-mail.

Shortly after a seller tells First Virtual that you bought information, First Virtual will forward a copy of the bill to your electronic mailbox for your inspection. The bill will include the seller's name and your name, the amount you are requested to pay, and a brief description of what you purchased. In addition, there may be a longer and more detailed description provided by the seller. If you are using First Virtual's electronic mail extension software, a convenient form explaining the details will appear, allowing you to reply automatically. If you are using your regular e-mail software, you should use the "reply" feature of your e-mail software, taking care to preserve the transaction code number in the header, and including in your reply message just a single-word answer. This single word must be one of these three answers: "yes," "no," or "fraud."

Whenever First Virtual posts a charge to your credit card, they will again send you electronic mail detailing the information purchases that make up that charge. This message will include a short code number that will also appear on your credit card statement. You can use this message, along with your credit card statement, to understand exactly what each First Virtual charge to your credit card was for.

When your credit card expires, First Virtual will send you a message instructing you to call its toll-free 800 number again to provide updated credit card information. You will be charged $2.00, which pays for part of the cost of updating your credit card information. If you need to change your credit card information at any time, send an e-mail message to initchg@card.com for an e-mail form and instructions. You must put your First Virtual account identifier in the subject line of your message. You can also use Telnet to connect to telnet.fv.com and select "Change" from the menu.

CommerceNet

CommerceNet is open to public and private organizations that subscribe to CommerceNet's charter to develop, maintain, and endorse an Internet-based infrastructure for electronic commerce in business-to-business applications.

CommerceNet offers two membership levels: sponsoring member and associate member.

The sponsoring member is a full member of the consortium and participates in both the governance of the membership organization and the direction of CommerceNet standards and technology. Sponsoring membership is open to any U.S. public or private organization.

Getting Involved

Sponsoring members' benefits include

- Membership on the Sponsor Steering Committee, which sets consortium membership guidelines and technology/standards direction.

- Voting membership in CommerceNet working groups whose charters are to define standards, discuss issues and requirements, provide guidance for CommerceNet deployment, and field pilot applications.

- Inclusion of the member's logo and name in CommerceNet's directory, providing an active hyperlink to the member's Internet World Wide Web server.

- Free access to CommerceNet basic and extended provider starter kits and user starter kits.

- Early access to CommerceNet technologies, with opportunities to serve as beta-sites for new software.

- Access to S-HTTP server and client reference implementations, with license agreement with Enterprise Integration Technologies.

- Eight hours of technical support from the CommerceNet staff for setting up an Internet storefront, if required, in the first six months of participation.

- Credits for four people to attend a CommerceNet-sponsored training class at no charge.

- Access to CommerceNet training materials for use within sponsoring member's organization.

Sponsoring members' opportunities for joint marketing with CommerceNet include

- Placement of CommerceNet's logo on products and services included in joint marketing programs.

- Inclusion of the member's promotional material with CommerceNet's press releases and collateral materials.

- Mutual referrals and leads.

- Presence during public presentations and demonstrations.

- Inclusion of sponsoring-member quotes and case studies in CommerceNet collateral and presentations.

- Joint participation in industry conferences and trade shows, with possible speaking opportunities for sponsoring members to represent CommerceNet.

- Collaboration with other CommerceNet members on electronic commerce issues, with the ability to influence and cooperate in the development of technology and standards for electronic commerce on the Internet.

The associate member is a member of the consortium and participates in the CommerceNet working groups. An international associate membership is also available. Associate members' benefits include

- Nonvoting membership in CommerceNet working groups whose charters are to define standards, discuss issues and requirements, provide guidance for CommerceNet deployment, and field pilot applications.

- Inclusion of the member's logo and name in CommerceNet's directory, providing an active hyperlink to the member's Internet World Wide Web server.

- Free access to CommerceNet basic provider starter kits and user starter kits.

- Access to S-HTTP client reference implementation, with license agreement with Enterprise Integration Technologies.

- Four hours of technical support from the CommerceNet staff for setting up an Internet storefront, if required, in the first six months of participation.

- Credits for two people to attend a CommerceNet-sponsored training class at no charge.

Associate members' opportunities for joint marketing with CommerceNet include

- Inclusion of the member's promotional material with CommerceNet's press releases and collateral.

Getting Involved

- Mutual referrals and leads.

- Presence during public presentations and demonstrations.

- Collaboration with other CommerceNet members on electronic commerce issues, with ability to influence and cooperate in the development of technology and standards for electronic commerce on the Internet.

The associate members elect four voting representatives to the sponsor steering committee to ensure that any specific issues are communicated and considered in the membership governance.

It is the responsibility of members to maintain the server connecting the member's organization to the Internet and to develop the home page and other material that are the member's Internet presence. The CommerceNet "Getting Started" class covers basic information on how to accomplish this.

For international associate members, U.S. export restrictions may apply to CommerceNet-licensed technologies.

To participate in CommerceNet as a provider you need an Internet connection, which can be obtained from an Internet provider.

Typical Internet storefronts require connection speeds of 56 kilobits per second (kbps) or higher. More popular sites are connected to the Internet at T-1 (1.544 Mbps) data rates. Check with your Internet service provider to make sure that your Internet connection will allow your organization to run a World Wide Web (WWW) client (e.g., Mosaic, to explore information and service offerings on the Internet) and server (to provide information and services to people on the Internet). A Unix system is recommended if you want to take advantage of CommerceNet software technologies.

You can start the registration process with CommerceNet by sending electronic mail to info@commerce.net or by calling (415) 617-8790. If you already have a WWW client, you can initiate the registration process online by filling out the CommerceNet information form available at the organization's Web site. An information package will be sent detailing the membership benefits and requirements.

If you do not have a WWW client, you can obtain a CommerceNet starter kit via anonymous ftp from ftp.commerce.net. The user starter kit is freely available to all.

Once you are a CommerceNet provider, you can design and implement online electronic catalogs, online ordering, procurement, and collaborative engineering data interchange services. The WWW Server can also be a gateway to your existing business systems. If you wish to outsource systems integration work, consult with the CommerceNet value-added directories or contact the CommerceNet Task Groups.

When you are ready with your Internet storefront (e.g., home page), announce your organization's presence on the Internet by linking your storefront to the CommerceNet directory. Send mail to webmaster@commerce.net and specify the uniform resource locator (URL) you wish the CommerceNet directory to reference for your Internet storefront. People with Internet access can reach your storefront by consulting the CommerceNet directories.

CommerceNet also offers a subscriber program that allows companies to easily and inexpensively create Internet presences or "storefronts." Although originally designed for small businesses, it is open to any organization that would like to participate in CommerceNet.

Benefits of enrollment include

- A listing in the CommerceNet Subscriber Directory, with a hyperlink to the subscriber's home page.

- A keyword-based indexing facility.

- A basic HTML starter kit.

- An online bibliography/summary of Internet business guides and publications.

Subscriber enrollment costs $400 a year for U.S. companies and U.S. subsidiaries, and $800 a year for non-U.S. organizations. There is a one-time $250 initiation fee.

The subscriber program also offers an optional one-day tutorial to help companies better understand how to construct and maintain an Internet presence. It costs $250 per person per class and covers areas such as

- Internet and CommerceNet overview.

- The World Wide Web and Mosaic.

- HTML tutorial.

- Web design basics.

To learn more about the CommerceNet subscriber program and CommerceNet's activities, send electronic mail to subscriber-info@com-

Getting Involved

merce.net, call the CommerceNet administrator at (415) 617-8790, or send mail to CommerceNet, 800 El Camino Real, Menlo Park, CA 94025.

In addition, CommerceNet offers a hosting service option, which helps businesses establish and maintain Internet storefronts. The component has a $600 annual fee and includes

- A megabyte of disk space on a server connected to the Internet via a well-maintained line with a minimum bandwidth of 56 kilobits per second.

- FTP or Telnet access to a hosting service computer (for subscribers to develop and update Internet materials as needed).

- Technical administration of the hosting server.

- Access reports (sent via electronic mail at least monthly).

Another business-oriented program run by CommerceNet is its Small Business Evangelist Program. The mission of this program is to increase CommerceNet's ability to attract businesses to its subscriber program by creating and motivating a group of "Small Business Evangelists."

CommerceNet has established a goal of having several hundred companies become subscribers by the end of its second year (April 1996). To help achieve this objective, CommerceNet will be giving commissions to the Small Business Evangelists based on the number of subscribers they attract to the program. Specifically, for each new organization an Evangelist helps become a subscriber, CommerceNet will pay a onetime commission of $200, given that the Evangelist signs up at least four subscribers. For example, if an Evangelist encourages six companies to become subscribers, then the CommerceNet pays a commission of $1,200.

This program will not only assist the evangelists financially, but it will also help them better pursue Internet-related business opportunities. The subscriber program will give internet service providers and Web consultants, for example, an attractive offering to supplement their own service package.

Evangelists have to meet the following conditions:

- Must be in business for at least six months.

- Must have a minimum of four customers.

> **CommerceNet has established a goal of having several hundred companies become subscribers by the end of CommerceNet's second year (April 1996).**

- Must provide CommerceNet with two customer references.

- Must provide CommerceNet with two character references.

- Must provide CommerceNet with a credit reference.

- Must be knowledgeable about the Internet and WWW.

- Must be able to consistently and accurately answer a series of background questions about CommerceNet and the subscriber program.

CommerceNet will regularly audit the performance of the evangelists by asking new subscribers a series of questions to gauge the evangelists' quality and service.

To obtain more information about the Small Business Evangelist program contact Kim Kasper at kimk@commerce.net or (415) 617-8790.

CyberCash

CyberCash's Secure Internet Payment Service allows merchants to operate an entirely automated "store" on the Internet.

CyberCash enables Internet merchants to safely process credit card transactions over the Internet and works with a merchant's bank to make possible immediate online authorizations.

The Secure Internet Payment Service includes all the functions needed to process transactions and settle them with the bank or processor. In many cases, credit card funds are available to a merchant the next day.

Some Internet merchants envision offering products and services at very low prices. One feature of CyberCash's Money Payments Service will be MiniPayments, which satisfies this Internet-specific need.

There are two criteria that every merchant must meet before using the CyberCash Secure Internet Payment Service for credit cards, for which merchants are not charged.

- A merchant must be authorized to accept credit cards by a bank or other agent.

- A merchant on the Internet needs to have a "store." The store runs on a server, which can be totally dedicated to one store or be shared among a few stores. Companies that help build stores on the Internet are called providers. They range from consultants to mall operators. You must have a complete, ready-to-run store for the CyberCash Secure Internet Payment Service to work.

Getting Involved

If you meet the above criteria, contact your credit card bank or agent and tell them that you want to use the CyberCash Secure Internet Payment Service. You may also contact any of the credit card banks or agents already offering the service and fill out an inquiry form. Send e-mail correspondence to merchant@cybercash.com.

CyberCash recommends a number of value-added integration providers (VIPs) to merchants who wish to set up shop with the company. These companies are

- QuakeNet, which can provide for all Internet needs. They were the first site to support multiple CyberCash merchants. E-mail: admin@Quake.Net.

- The Electronic Pen focuses primarily on conceptualizing, designing, and building interactive Web sites along with server coordination and continued maintenance. E-mail: webcats@epen.com.

- Internet Café designs and hosts WWW documents, in addition to providing dedicated and dialup Internet access in Santa Barbara, California. E-mail: rad@internet-cafe.com.

- Potomac Interactive Corporation consists of more than fifteen professionals dedicated to World Wide Web presence design, programming, and strategic planning for a wide variety of companies and organizations. E-mail: hankd@picnet.com.

- Lokian provides for the delivery and coordination of all aspects of a WWW presence. The company provides a comprehensive approach to addressing and resolving strategic and tactical business, marketing, competitive, and technology issues, fast. E-mail: info@lokian.com

> **CyberCash enables Internet merchants to safely process credit card transactions over the Internet and works with a merchant's bank to make possible immediate online authorizations.**

DigiCash

To use Ecash you need a computer connected to the Internet, the Ecash client software, and an Ecash account.

You can get the Ecash client software for free by downloading it from the DigiCash Web server (http://www.digicash.com/). It is a small application that runs next to your Web browser. It lets you withdraw Ecash from your account and spend it at any time while surfing the Internet, just as you withdraw and spend cash outside cyberspace.

As part of an Ecash trial, by registering you can open an Ecash account with no costs or obligations. During the trial, DigiCash will put $100 in your account, for free.

When you register you will get a password that allows you to download the software and open an Ecash account. This password will be sent to you via e-mail within a couple of days. When you have received your password go to the Ecash client software page on DigiCash's Web server to download the software. Your account will be opened automatically when you run the software for the first time.

Registration information will not be made available to any other party.

DigiCash calls its storefronts "cybershops." A cybershop consists mainly of two parts: one that handles the files, data, or access that you want to sell, and another that handles payments. The first part is basically a Web server that allows access restrictions. The second part sends payment requests and processes the responses.

There are two ways to handle the money collections. One is to build it yourself from the Ecash software and some shell scripts. Examples of this method are provided at the DigiCash Web site.

The other way is to use DigiCash's Remote Shop Server. This is ideal for people who run a Web server but cannot run an Ecash shop on their machine (either because their platform is not supported or because they are not allowed to run scripts).

The money collected by cybershop merchants is Cyberbucks, the money issued by DigiCash in the Ecash trial. The Ecash that is paid to a shop goes directly to the merchant's Ecash account at the DigiCash's digital bank. A merchant can withdraw his Ecash at any time.

Starting an Ecash-accepting shop on the World Wide Web is almost as easy as installing the client software. If you have a Web server running and you have installed the Ecash client software, you can use ready-made script files to start your shop and start asking Ecash for any data or access you wish to sell. The Ecash paid to your shop goes directly to your Ecash account at the digital bank.

If you build Web pages, but are not familiar at all with scripts and gateways, you still can start a shop with DigiCash's Remote Shop Server.

Open Market

Open Market provides a number of services for a variety of business types.

For publishers, Open Market can

- Convert a proprietary online information service into an open pay-per-page service targeted at specific customer segments.

> **As part of an e-cash trial, by registering you can open an e-cash account with no costs or obligations. During the trial, DigiCash will put $100 in your account, for free.**

Getting Involved

- Convert a print-based publication and marketing expertise into online services focused on defined interest areas.

- Implement a service that enables publishers to get paid for corporate, professional, and individual copies of copyrighted publications.

- Create a service that includes the automatic fingerprinting of reports to protect the publisher against unauthorized use.

For banking and financial services companies, Open Market can

- Create an online community of bank customers in a region.

- Create the infrastructure and security required to conduct large financial transactions on the Internet.

- Create new financial mechanisms for handling corporate purchases over the Internet.

- Create the infrastructure to form a community of independent businesses centered around a major bank.

For advertising firms, Open Market can

- Create the infrastructure for combining the reach of newspapers with the information-provision capabilities of the Internet.

- Create the tools and business approaches to integrate print advertising with network-based advertising, taking advantage of the unique properties of each medium.

- Create the systems to track detailed information about users (not available through standard Web protocols) to define advertising rates on the Internet.

For remote retailers, Open Market can

- Create an infrastructure to offer small customers customized services previously available, because of their cost, only to large customers.

- Automatically create small interactive stores that have complete payment and management capabilities.

For industrial clients, Open Market can

- Create the capability to deliver coupons for the purchase of items through traditional channels.

- Create the infrastructure to do controlled market research through surveys over the Internet.

APPENDIX

An Interview with Tim Jones, CEO of Mondex

Tim Jones is CEO of Mondex. He is the coinventor of the "electronic cash" smart card and one of fifteen industrialists advising the British Trade and Technology Minister, Ian Taylor, on multimedia and information highways.

What do you see as Mondex's role in developing the environment for electronic cash?
I see Mondex as being one of the infrastructure pieces of electronic commerce. If you go back a couple of hundred years, you have the world of physical commerce setting down its infrastructure. Because physical comerce is physical, that was about roads and railways and canals, and later on, it was about telex and telephone services.

What's happening in electronic commerce at the moment is exactly the same thing conceptually—but, of course, the infrastructures are different. The single biggest difference between electronic commerce and physical commerce is that the priority of geography largely disappears in electronic commerce. For the first time, it doesn't matter where you are. What binds people together now is common interest rather than common location.

If you look at activity in physical commerce, the priority of geography was very important. Where you were determined the markets you could serve easily, the raw materials and human resources available to you. Much of that priority begins to disappear with electronic commerce. If you want to work in a particular way with a particular set of counterparties, more and more you're going to be able to find them on electronic networks wherever they are and work in that way.

The fundamental thing about the infrastructures for electronic commerce is that they're global; they're not territorial. If you put them up on a territorial basis, you miss the point.

If you look at what those key infrastructures are, some pieces which were already already very powerful in physical commerce moved directly across to electronic commerce—telephones, PCs. Other things in my business, the payments business, have also moved across. I expect Visa and MasterCard to be very important infrastructure brands in electronic commerce, performing roughly the same roles that they do today in physical commerce: allowing for individual transactions to be undertaken, typically between consumers and merchants, but I think they will play an important role in individual business-to-business payments as things evolve.

Mondex fits into this because Mondex is cash on the Net. It shares an important feature with Visa and MasterCard in that it is not just for the Net. I don't believe that consumers will tolerate payment options that are Net-specific when they have other payment options available to them which are universal. MasterCard and Visa will succeed because you can use them on High Street and when you get home, you can use them on the Net. Mondex will succeed for the same reasons. It is just as happy in the physical world, in McDonald's, in fast food retailing, as it is across the Internet for shareware at a Web site.

I see Internet payment options as short- to medium-term things that will die away as systems that support more ubiquitous payment options come into play.

What's the time frame for the development of electronic cash?
Most of the infrastructure problems for payments will be solved by the end of 1997. But the rollout of systems will take another decade beyond that.

To get parochial for a minute, for Mondex there are no new counterfeiting issues because we don't depend on the Net for any component of security. The Net does pose a challenge for us, a diversion challenge. A hacker could notice that someone was about to pay with Mondex over the Net and could divert the money to his purse so you'd be paying someone you didn't want to rather than paying the person you thought you were paying. Then you wouldn't get the goods and you'd reconnect to the Web site and say, "Look, I sent you my money. Where's my shareware?" and the Web site would say, "No, you didn't."

I thought you had solved the diversion problem.
There are various levels of sophistication in a diversion attack. It's extremely unlikely that you're going to get a successful one in physical commerce. In electronic commerce, it depends on how sophisticated the hacker is.

Suppose you're displaying the purse narrative to the screen of a counterparty. You're telling the counterparty that this is my purse's narrative; it can only be my purse's narrative because my purse says so. If the hacker is clever enough to invade your client, or counterparty, and intercept the purse narrative sent to you, he can substitute his I.D. for the counterparty's. So you will be looking at the counterparty's narrative but send your money to the hacker's purse. You'd find out later because when you audited your card, you'd find out that the counterparty, rather than being World Wide Web Software or something, would be Ha Ha Ha or some such.

These aren't insurmountable problems. We expect to have a solution by the end of 1995, and we expect to be doing things on the Net in 1996 with a full rollout in 1997.

Is it accurate to refer to your system as an "unaccounted" system?
It depends on how you define accounting. Mondex is a sophisticated system with several classes of purses. Each class has different value limits and different business rules that apply to them.

If you talk about the high-value purses that will be in member banks and in large retailers, they will effectively be accounted for. Those purse holders will be required to report what they do in some way.

When you go to the consumer purses, they will have relatively low limits, $1000 or £500. Those purses remember what they've done for the last ten transactions. That may be increased in the future. There's nothing in the architecture of the purse to limit where we set that memory. As flash memory technology improves and its price goes down, we may want to increase that number to 1,000.

This "unaccountability" is important for keeping the cost of transactions down, isn't it?

Absolutely vital. And it's also important for balancing the state's legitimate need for oversight with the consumer's legitimate desire for anonymity.

If we look at the economics, the marginal value of sending Mondex value across the Net if you're already connected is vanishingly small. It's a small amount of data, a couple of messages and you've done it. It's entirely viable to think of shareware being transformed as a market by Mondex. You'll be able to have Web sites popping up all over the place that have specializations in shareware—fonts, images, bits of software. They can rent a Web page for a couple of dollars a week or whatever and say, This font is 50 cents, that one a dollar. At the receiving end, there'll be a purse. All that purse does is collect money. It doesn't have to do anything with it. There's no fixed cost for putting it through a clearing system. You can have separate purses for salespeople and software writers. When a sale is made, the money goes into the separate purses. Then the artist or salesperson can phone the Web site and transfer the money to his purse at home. That's all economically viable. But as soon as you start accounting for the transactions, you start running into fixed transaction costs that make those levels of transactions uneconomic.

The other thing is, a product that accounts for these transactions isn't very exciting because it isn't money. If I'm dealing with someone face to face and I want to exchange something called money with him, I'd expect to give you some and I'd expect to be able to receive some from you. Exactly the same should apply to electronic commerce. Just because you're 3,000 miles away doesn't mean that I shouldn't be able to send you money or receive money from you. If we were talking on Mondex phones, I could pop my Mondex card in my phone and you could pop your Mondex card in your phone and I could send you money. That's how money works. There's a deep fundamental about what money is, which I believe Mondex obeys and an accounted-purse architecture does not. One is a preauthorized debit card and the other really is money.

Think what that means to the banking sector. If you have a phone or a PC with the appropriate software, I can say to you in the future, Put Mondex on that and I can turn it into a cash dispenser. That's a defensible statement by me because what comes out of that device onto your Mondex card is cash. You can give it to members of your family, you can give it to your friends, and you can receive value on it from them. It obeys the rule of transferability, which makes money the product that it is. It's different from saying, I'll give you this PC upgrade pack and you can put value onto a card which you can use with participating merchants. I can do that with a debit card without having to get the value ahead of time. That may be mildly interesting, but it's not money.

As the dust settles in this market, the differences between Mondex and accounted purses will be better understood. I'm not saying that accounted purses won't be successful. What I'm saying is, they don't do what Mondex does, and Mondex will be successful.

How can electronic money reduce revenues to a government?
Banks have to buy cash from a government. Mondex competes with that. It takes market share from the government. Today the beneficiary of cash is the government. Tomorrow it will be the stockholders in the originators of Mondex value. What I think will happen, though, is the government will just tax it back. So we say to people, don't expect to hold on to the money you make from gaining market share, because you won't. Governments aren't stupid. As soon as they see what's happening, they'll get it back in some way.

Remember, when banknotes came out they, too, were a private-sector initiative that was eventually nationalized. So we're not doing anything novel here.

Do you foresee the government nationalizing Mondex?
They could nationalize the origination of the value without nationalizing the brand. Our corporate structure provides for central banks to act as originators. We did that on purpose. It's part of the design.

Presenting Digital Cash

An Interview with David Chaum of DigiCash

David Chaum is managing director of DigiCash, a firm on the leading edge of digital cash development. He is chairman of the European Union Project CAFE, which combines smart cards and software-only electronic money. He has published over forty-five original technical articles on cryptology and developed a method of ensuring an individual's privacy in digital cash transactions. We spoke to Chaum at DigiCash's headquarters in Amsterdam. At ten P.M. he had just finished his Chinese take-out and was winding down after another long day.

What is DigiCash's role in the e-cash arena?
We're an R & D company. We make software to do this and some chip cards and we will license our technology.

You have been characterized as one of the leading advocates for privacy in the digital cash area. How do you feel about that?
I'm not happy about being pigeonholed that way, but it's something that has happened.

Is privacy necessary for the public acceptance of digital cash?
I'm not sure about acceptance. But in the medium and long term it will be an increasingly important feature of electronic cash. Some would say, and I would agree with them, that it really is not electronic cash unless your privacy is protected. That's the idea of cash: You don't have to reveal your identity to pay.

Cash is the most popular payment system on the planet. It's something even kids know how to use. Its particular structure may have had one technological origin, but it's now established as a fundamental payment system and certainly the most prevalent one. If you want to create an electronic thing that will replace cash, at least in cyberspace, then it should have all the features that cash has. It can have more features, but it can't have fewer.

People in cyberspace recognize the power of information much more than people outside cyberspace. The first wave of awareness about privacy arose when people began seeing their names on computer punch cards and printouts. They began to get a glimpse of how computers could keep track of them. When people are exposed to cyberspace, you can expect a second wave of privacy concerns.

Appendix

Is there something about electronic transactions that make them more of a threat to an individual's privacy?
The granularity of the information about a person's activities and interests that's revealed in electronic transactions is far higher than the amount of detail which could be gleaned from the payments people make today. Those payments are much coarser than electronic payments.

As people see the power of information and micropayments become more prevalent, this will work to make people want privacy in electronic payments. People will have a lot of choice in things like electronic payments, and providers will realize that. The whole market share of Netscape went from 10 percent to 80 percent in ten days. People find it very easy to change things on the Internet. If people want this stuff, they will have it on the Internet. And it's clear they will want it more and more.

Do some of the electronic cash systems coming online now, like Mondex, adequately protect privacy?
I love Mondex because it protects privacy. They're private in practice, although they're not as nice as some e-cash payments, in that the person's computer doesn't prevent other computers from learning its identity, but payment data aren't made available to central sites.

Do you agree with Mondex that electronic cash systems need to be universal in order to be viable?
Mondex has merit because a payment system becomes more useful to people the more universal it is. But there are all kinds of specialized payment products that are making tons of money—traveler's checks, money orders—and they're all happily coexisting. They're accessible to different kinds of people and they're useful for different amounts of value. So I think it's silly to say that.

It'd be nice to have one universal system, but the history of payment technology has been that every payment system ever invented is still in use. They just lose market share.

Privacy could become a universal feature, though. Once people become aware of privacy and start to use it, I think they'll be offended if it's not offered to them in other contexts as well, because they'll know that such a mechanism is possible.

Today people believe that anything to do with computers is traceable. Once that is shattered, people will more than insist on privacy, they will expect it. It will become an informational human right. It's just a continuing of the trend of empowerment that has made the PC what it is.

There have been studies that conclude that the best way to control the American population is to outlaw banknotes. It's not that implausible that under certain circumstances paper money would be outlawed. Then you would have only electronic forms of payment and one could predominate easily.

I agree that things would be better if all cash were electronic. It's superior to paper money and could solve a whole lot of problems.

Won't a totally electronic system place too much power in the hands of the state?
Not necessarily. It's pretty hard to stop people from doing things on the Internet. Once they start moving bits around, you can't really tell what they're up to.

My impression is that banks feel a little uncomfortable with anonymous payment schemes. Is that so?
I prefer not to call it anonymity any longer. I call it one-way privacy. That's because the person who makes the payment can always allow the payment to be traced. The person who receives the payment can't find out the identity of the person who paid without the cooperation of the person who paid. It isn't like two people with paper bags over their heads meeting in the middle of nowhere.

Most banks that I've spoken with are quite positive about it. Banks aren't really interested in low-value payments or keeping a lot of transaction data about them. It's a nuisance. Then you have the problem of chargebacks, where banks have to give refunds and settle disputes. Banks don't want to have to do that for 25-cent payments. They can't afford to do it. Banks like to know their customers, but not at this level of detail.

Another thing is the image of retail banking. One reason banks have access to consumer capital is they protect the confidentiality of their client's financial transactions. So it would only be natural that banks embrace something that would help protect financial confidentiality. If they disregarded it, it would be harmful to their image. Banks that offer anonymity will have a competitive advantage over banks that do not.

Appendix

How fast do you think electronic cash will develop?
E-cash is just software. You just download it from the Net and run it. The infrastructure is there for that. That differs from smart-card systems where you have to get cards into everyone's hands. That costs money and takes time. Then you've got to get all these card-accepting devices at points of payment. That's a chicken-and-egg problem. Retailers won't be willing to pay for that stuff unless consumers have it, and consumers won't want it unless they can spend their e-cash in a lot of places. All that is going to take quite some time. And it might even fail. But on the Internet, it will grow as the Internet grows.

An Interview With Dan Schutzer of Citibank

Dan Schutzer is vice president for advanced technology at Citbank in New York City. He has also authored a "white paper" on digital cash which is posted at the Web site for the Cross-Industry Working Team, a multi-industry coalition committed to defining the architecture and key technical requirements for a powerful and sustainable national information infrastructure. The white paper can be found at http://www.cnri.reston.va.us:3000/XIWT/documents/dig_cash_doc/ToC.html.

What do you think are the major barriers to the development of electronic cash?
First, there's a lack of standards. Real cash is accepted anywhere. Digital cash comes in a variety of forms and implementations. There's DigiCash, Mondex, Smart Cash and others. They're all different. It's not clear the same readers will work with them all. It's not clear that they're all worth the same amount of money. If I step up to a vending machine with a card full of this stuff, is it going to work? And if it doesn't work universally, it doesn't seem to me that it's going to be too useful.

Can this be overcome? Sure it can. It was a barrier overcome by credit cards and ATM cards. But it's a more formidable barrier here because we're talking about cash. We have to feel comfortable that it's not counterfeit. We have to be able to use it without feeling the frustration that in most places we go into it will not be accepted because people are jittery about it.

One problem with digital cash is it's not a precise term. It depends on the implementation. Some implementations allow for peer to peer transfers. Others require the cash to returned to the bank as soon as it's spent with a merchant.

There are also questions about who's backing it. Is it the U.S. government? Is it a regulated bank? Is it Microsoft? Is it Joe's Diner?

Do you think digital cash can be a replacement for debit cards, checks, and money orders?
It could be, but its most obvious replacement is for cash. Cash is expensive to handle. You've got to worry about transporting it. That means armed guards. You've got to store it some place safe. That's expensive. Digital cash is easier, but then you have to worry about computer crashes wiping out your cash.

Appendix

It costs more than a dollar to handle a cash transaction. So it is very attractive to a merchant not to have to handle cash.

It's attractive to consumers, too. It's perceived as convenient, not as heavy to carry, faster because you don't have to wait for people to count it.

So it's likely digital cash will cut into the use of hard cash. It's less likely it will cut into credit cards because credit cards offer consumers different features—30-day float, the ability to spend beyond your means, ties to things like frequent-flier miles. On the other hand, debit cards, where you are physically drawing against cash, might be pinched by digital cash. But debit cards give you a lot more transaction detail than some implementations of digital cash, so if I were a business or someone who really had to keep track of what I was spending, debit cards would still be attractive to me.

Do you think digital cash will provide a rationale for banning hard cash entirely?
No. Cash is going to be here forever. There are business reasons why merchants and individuals want to move away from cash. Simply, because it's a pain in the butt.

But digital tokens aren't necessarily the answer. People could find it more convenient to walk around with a smart card that was universal and secure. Anytime they want to perform a transaction, credit or debit, they could do so with the card. If I walked into your store and wanted to buy something on credit, the card would act as a credit card. If I wanted to debit the purchase, the card would do that for me, too. And if I wanted to transfer funds from one of my accounts to your account, the card would also do that for me.

How important is anonymity to the acceptance of digital cash?
I think it's a red herring, but it's a very emotional thing. The average person uses a debit card or a credit card in a variety of transactions that would surprise you. They'll go to a hotel and watch an X-rated movie and charge it to their credit card. They don't care. They'll go a store and use a store card. That card tells the store what you buy. People still do it. So to me, it's a blown up thing. You have a lot of civil libertarians making a lot of noise but they don't represent most of the public. That's my personal opinion.

People want convenience and ease of use. And they want to keep track of what they spend their money on. They can't do that if the transaction is anonymous.

Now if we start talking about the illegal gardener, the illegal alien, the money launderer—people like that—it might be important to them. But even they probably won't trust it. Are they going to learn logarithms so they can feel safe with this stuff? I don't think so.

People inherently think that if something is digital, it can be traced. You will have a difficult job persuading somebody that anything digital is truly anonymous.

I think there is a market for digital cash, but anonymity isn't a driving imperative behind it.

What is the role of the banking industry vis à vis digital cash?
We're looking at it because we deal in money and if it becomes popular, we want to be there. Most banks are reactive in this area, not proactive. We're waiting to see if there's any interest in it.

We're not pushing it because we don't think there's any money to be made on it. It's like cash. We don't make any money on cash. We'd be doing it as a service.

Don't banks make money on the float associated with digital cash?

I don't think banks will make much money on float. We have stored-value vehicles now, things like Traveler's checks, pre-paid phone cards and transit cards, which have some float. You might not use up all your Traveler's checks for a month or so, or value may stay on your phone card for weeks. But if the digital cash card was universal, people would keep very little value on their cards just like they keep very little cash on their person now.

Do banks have an advantage over non-banks in issuing digital cash because people trust banks more than they trust non-banks when it comes to money matters?
It depends how digital cash is introduced. If governments were to say, in addition to printing bills, we will generate these electronic tokens, and we will back them, then banks may not have any advantage over anyone else. But if an association of banks were to back a currency versus an association of Solomon Brothers, AT&T and Fidelity, then the banks might have an advantage.

The risk of using money issued by a non-bank is that when you went to convert the digital value into real value the issuer might not have the money to back up the digital cash. Whereas if a person stuck with the stuff issued by a Barclays' Bank, Citibank or a Central Bank, they would probably feel more confident the digital cash could be converted to hard cash.

Do you think digital cash will lead to the introduction of competing currencies?
That's a disaster looking to happen. That will allow companies, as they get more and more desperate, to issue more and more money without any backing. Eventually people would lose their life savings, there'll be a run on money, and it'll be 1929 all over again. That's why the government might be a little interested in who can issue cash.

An Interview with Don Gleason of the Smart Card Enterprise

Donald J. Gleason is president of the Smart Card Enterprise, a business unit of Electronic Payment Services in Wilmington, Delaware. In the summer of 1995 EPS formed SmartCash, a company whose mission is to develop, finance, implement and manage a national system for stored-value cards. Among the founding owners of SmartCash were Banc One, Bank of America, Chemical Bank, CoreStates Financial Corp., and MasterCard. SmartCash plans to roll out its stored-value product some time in 1996.

Is your company committed to electronic cash?
It's all we're focusing on right now. We've dedicated exclusive resources to it for three-and-a-half years.

Your stored value system is designed to work outside the Internet and exclusively at merchant locations. Do you anticipate expanding the system to the Internet?
Potentially, but it's not a focus right now. A lot of security issues would have to be addressed before we would be interested in doing anything like that.

Was security the reason you decided to exclude an Internet component from your system?
The primary reason we didn't address the Internet issue was that we decided to go after what we feel is a bigger market first.

What do you see as the major barriers to electronic cash?
The two major barriers as I see it are standards and education.

Is there any hope in getting standards in place?
Yes. Absolutely. We're actively involved in several groups that are developing and resolving standards issues. We're also working with Visa and MasterCard. Another issue relative to standards is critical mass, so we're forming strategic alliances to address that issue.

What kind of standards are needed?
Two kinds: technical and application. The technical standards are there—where the chip is placed on the card, the frequency it modulates at, those kinds of things. The primary focus on standards right now is in

the applications area—how stored value is treated by the card.

What's behind the stored value in your cards?
Money that the customer has deposited with his financial institution.

Are you doing anything with peer-to-peer transfer?
No. Not because we can't, but because we don't want to. There's no way to make money doing that that we can see, and we're a for-profit company. That's number one. Number two is that as a company, we're focused on the payments industry, primarily at merchants' locations, so we're staying within our core competency there. And a third reason is that if for some reason the system becomes compromised and you're doing person-to-person transactions, in our opinion there are no effective ways to audit those transactions so there are serious risks presented by that.

How do you make your money on electronic cash?
Primarily by merchants paying transaction fees, which are lower than credit card transactions fees, for purchases. It's also a flat fee regardless of the face amount of the purchase.

You see the average credit card transaction is between $45 and $55. We're targeting $20 and under transactions here. People aren't using credit cards for those transactions anyway. You don't use a credit card or write a check for a $10 purchase in most cases; you pull out a ten-dollar bill. So for purchases between five and twenty dollars, the prevalence of transactions are cash. And so there is no card-based solution for that that is cost-effective except this technology because it's off-line.

That doesn't mean that the terminal is off-line. At the end of the day all the electronic cash transactions are batched and sent to the bank online. That terminal may also accept other forms of payment that require online verification such as credit cards.

Do you think the stored-value card will replace other forms of payment such as debit cards, checks and money orders?
I think it will replace cash. I don't think it's going to replace anything else. I think it's going to be complementary to credit and online debit. All it's going to do is automate cash transactions primarily in the five to twenty dollar range.

What do you think of digital cash systems that build in anonymity and peer-to-peer transactions?
Those are entirely different systems from ours. We're not going after that market. I don't know what a company's business case is for doing those things.

How important is anonymity to electronic cash?
I think it's extremely important, but I think equally important is audibility. You have to construct a system which provides for both, which is what we've done. Neither the merchant nor ourselves would be able to tell who that consumer is without the consumer's bank telling us who a consumer is. The customer's identity remains anonymous unless it needs to be known by police or regulatory authorities because of a perpetration of fraud or an incursion into the system. The government is very interested in having monetary policy remain under control.

Do you think electronic cash competes with the government's issuance of cash?
I don't think it competes with the government at all.

Appendix

An Interview with Shikhar Ghosh of Open Market

Shikhar Ghosh is a co-founder and chief executive officer of Open Market.

What is electronic commerce?
When people deal with electronic commerce, they deal with it in a digital kind of way. The reality is that electronic commerce will be moving much more in an analog sort of way. Many functions that are done through physical means today are going to be done electronically.

The real growth in electronic commerce will be in hybrid forms. I spoke to a guy the other day who bought his car over the Internet. He did that because he knew exactly what he wanted. He went to all the dealers near him and got the best price he could get, which was $32,000. He went on to Autonet and got exactly the same car for $30,000.

Increasingly, what you're going to have is that the value of the electronic networks is going to be one of an intermediary function where someone is willing to perform a transaction—selling you a car, for instance—for the cost of doing the transaction. No physical retailer will be able to keep up with that.

Before the end of the decade, a very significant part of the economy is going to be negotiated electronically. I think a reasonable number is that about 37 percent of all business will be conducted over electronic networks by the year 2000.

The ability to search great quantities of information, the ability to not be bothered by geography, the ability to connect to any fulfillment house and the very low cost of doing all this are going to drive the major parts of your economy into the electronic modes of distribution.

What does Open Market do to fertilize electronic commerce?
In order for electronic commerce to occur you need an electronic analog of what happens outside electronic commerce. When you set up a business, you do a lot more than accept a credit card. You have a way to display your product. You have a way to get customers to come back. You have a way of managing customers. You have a way of doing a transaction. And you have ways of providing customer services. Only a narrow piece of that is payments. And only a narrow piece of payments is credit card transactions.

We do a spectrum of things. We do credit cards, but we can do billing, ACH [Automated Clearing House] transactions or create our own currency. You can use frequent flyer miles or service credits. We have created a whole infrastructure for electronic commerce and within that infrastructure, the form of payment is one slice of it.

We like to think of ourselves as one level above the form of payment. We accept credit cards but we also will accept cybercash, First Virtual payments or eventually e-cash. It just doesn't matter to us what the form of payment is.

We take a buyer and we take a seller and then we say, what form of payment will you accept from each other? For example, you might go to an online Apple computer store and they might accept Applebucks—service credits for what they've done before. That's a form of payment to us.

But credit card payments are the most common form of payment in electronic commerce right now?
Yes, for consumer transactions. When you go to these conferences on electronic commerce, what most speakers don't realize is that the credit card world defined through catalog business is about seventy or eighty billion dollars worth of business. If you take small transactions, under $350, between businesses, that's a $350 billion business, five times the size of catalogs. Most of that isn't done through credit cards. Most of that is done through purchase orders or ACH transactions.

Where will electronic cash have an impact on the system?
Electronic cash is extremely useful in two situations. One, in situations where you have the need for anonymity. Regular cash allows for that. I can go somewhere and buy a magazine and no one knows I did it. I can't do that if I use a credit card. I can do that if I use electronic cash. The other thing electronic cash is good for is very small transactions. If I'm doing transactions for a penny, a system that requires crediting someone and debiting someone make the cost of the transaction greater than the value of the transaction. Therefore, you can't do it. Electronic cash is very good for that.

Other than that, if you compare electronic cash to other payment instruments that exist, it's debatable as to how good it is or why people would use it. But if you want anonymous transactions, then there really isn't any other way to do it other than electronic cash.

I think that for several years the market for electronic cash is going to be quite limited. It's a very good substitute for a debit card. If you go to countries with a poor infrastructure, something like that may be just fine.

Appendix

Where do you think electronic cash will be more popular: on the network or off the network?
I don't know. If you have real cash on the network, you run into some pretty horrendous security problems. Anything on the network can be broken. Until that is resolved, you're not going to get big amounts of money being transferred this way. Off the network, you have a physical device—a smart card—to protect you.

People appeared to be getting comfortable with security on the Net. Has the problem with Netscape's software set back the development of electronic commerce on the Net?
I think it will, but I think it will blow over pretty quickly. Most people don't understand security on the network and from what happened, it appears Netscape doesn't understand very much either. What they did was a rookie error in implementation. They took a shortcut and got burned. It will be remedied quickly, but it really damages credibility. It's going to set the issue back by a few months but not much more than that.

What do you say to prospects when you try to sell them on Open Market and they point to the Netscape gaffe as a reason not to get involved in electronic commerce?
We tell them we don't use that technology, which is the reason they'd use us and not Netscape. Netscape works fine if you're doing a $100 transaction on a credit card. If you want to do a $10 million wire transfer, it's absolutely the wrong technology to use. It's the same thing as security for your house. A lock may be fine, but if you're storing gold in your house, it isn't. We can provide a continuum of security services down to Netscape-like technology.

Company Directory

Here is contact information for many of the companies discussed in this book.

CheckFree.
Phone: 614/825-3000
Address: CheckFree, P.O. Box 2168, Columbus, OH 43216
Website: http://www.checkfree.com/

CitiCorp.
Phone: 212/559-1000
Address: 399 Park Avenue, New York, NY 10043
Fax: 212/291-6633
Website: http://www.citicorp.com/

CommerceNet.
Phone: 415/617-8790
Address: CommerceNet, 800 El Camino Real, Menlo Park, CA 94025
E-mail: info@commerce.net
Website: http://www.commerce.net/

CyberCash.
Phone: 703/620-4200
Address: CyberCash, Inc., 2100 Reston Parkway, Reston, VA 22091
Fax: 703/620-4215
E-mail: info@cybercash.com
Website: http://www.cybercash.com/

DigiCash–Headquarters.
Phone: 011/31/20-665-2611
Address: DigiCash bv, Kruislaan 419, 1098 VA Amsterdam, The Netherlands
Fax: 011/31/20-668-5486
E-mail: info@digicash.nl

DigiCash–U.S.A.
Phone: 212/909-4092, 800/410-32274
Address: DigiCash, 55 East 52nd Street, 27th floor, New York, NY 10055

Fax: 212/318-1222
E-mail: office.ny@digicash.com
Website: http://www.digicash.com/

Electronic Payment Services (SmartCash)
Phone: 302/791-8200
Address: 1100 Carr Road, Wilmington, DE 19809

First Virtual Holdings.
Phone: 619/793-2700
Address: 11975 El Camino Real, Suite 300, San Diego, CA 92130
Fax: 619/793-2950
E-mail: info@fv.com
Website: http://www.fv.com/

MasterCard.
Website: http://www.mastercard.com/

MCI Communications Corporation.
Phone: 202/872-1600
Address: 1133 Nineteenth Street NW, Washington, DC 20036
Fax: 202/887-2154
Website: http://www2.pcy.mci.net/marketplace/index.html

Mondex.
Fax: 011/44/0/171-920-5505
Address: 1st Floor, Podium, Drapers Gardens, 12 Throgmorton Avenue, London, EC2N 2DL, England
E-mail: news@int.mondex.com
Website: http://www.mondex.com/mondex/home.htm

NetCash.
Phone: 213/743-4926
Address: Keith Johnson, Office of Patent and Copyright Administration, University of Southern California, 3716 South Hope Street, #113, Los Angeles, CA 90007-4344
Fax: 213/744-1832
E-mail: kjohnson@opca.usc.edu
Website: http://nii-server.isi.edu:80/gost-group/products/netcash/

NetMarket.
Phone: 617/441-5050, 800/867-3777
Address: The American Twine Building, 155 Second Street, Cambridge, MA 02141–2125
Fax: 617/441-5099
E-mail: info@netmarket.com
FTP: ftp.netmarket.com
Telnet: netmarket.com
Website: http://netmarket.com/

NetScape.
Phone: 415/528-2555
Address: 501 E. Middlefield Rd., Mountain View, CA 94043
Fax: 415/528-4124
E-mail: info@netscape.com
Website: http://www.netscape.com/

Open Market.
Fax: 617/621-1703
Address: Open Market, Inc., 245 First Street, Cambridge, MA 02142
E-mail: feedback@openmarket.com
Website: http://www.openmarket.com/

Security First Network Bank.
Address: 300 Virginia Avenue, P.O. Box 7, Pineville, KY 40977
E-mail: comments@sfnb.com
Website: http://www.sfnb.com/

ViaCrypt.
Phone: 602/944-0773
Address: 9033 North 24th Avenue, Suite 7, Phoenix, AZ 85021
Fax: 602/943-2601
E-mail: info@viacrypt.com
Website: http://www.viacrypt.com/index.html

Visa.
Website: http://www.visa.com/visa/

INDEX

A

accessing Internet, 26
accounts
    CyberCash, 118
    First Virtual, 197
advance payments (First Virtual), 202
Advance Publications (Open Market), 151
advertising, 38
algorithms (digital signatures), 49
American Bankers Association, 177
ARPANET, 22
    Crockett, Dr. Stephen, 116
AT&T (virtual mall), 69
ATMs (banks), 178
Automating transactions
    (First Virtual), 204

B

Baker, Stewart, 53
Bank One (Open Market), 148-150, 157
banks, 177
    American Bankers Association, 177
    ATMs, 178
    branches, 178
    CyberCash, 114
        First National Bank of Omaha, 124
        Wells Fargo & Company, 124
    DigiCash, 112
    ecash, 178-180
    First Union, 156
        Open Market, 146
    Mondex, 97
    redlining, 179
    Security First Network Bank, 242
baseball (marketplaceMCI), 71
Bermanis, Jerry, 45
billing (First Virtual), 131
blind signatures, 102
    Chaum, Dr. David, 99
    privacy, 100
    security, 100
    transactions, 101
Borenstein, Nathaniel (First Virtual), 135
Boston Camera View (Open Market), 138
browsers
    Mosaic (NetMarket), 173
    Netscape, 171
buyers (First Virtual), 206
    see also consumers

C

CAFE (Conditional Access for
    Europe), 107
Cerf, Vinton, 34
Chaum, Dr. David, 99
    DigiCash, 99, 103
    interview, 226
CheckFree, 163
    address, 240
    WWW site, 240
CheckFree (CyberCash), 123
checks
    Federal Reserve, 178
    forgers, 178
    NetCheque, 168
children (Mondex), 89
choice (mass marketing), 37
Citibank (Dan Schutzer), 230
CitiCorp
    address, 240
    WWW site, 240
Clark, Dr. James H. (Netscape), 171
class issues, 183
Clickshare, 191-193
clients (First Virtual), 136
Clipper Chip, 51
    backdoor, 53
    Baker, Stewart, 53
    Gore, Al, 57
    key escrow encryption, 55
    NSA, 56

privacy issues, 53
private sector, 55
secrecy, 56
Skipjack, 53
wiretaps, 53
collections (First Virtual), 131
CommerceNet, 165-167, 210
   address, 240
   fees, 214
   membership, 210, 213
   registration, 213
   Small Business Evangelist Program, 215
   starter kit, 213
   storefronts, 214
   WWW site, 240
Community-Commerce (Open Market), 147
complaints (consumers), 41
CompuServe
   S-HTTP (Secure Hypertext Transfer Protocol), 76
   virtual mall, 69
Condé Nast (Open Market), 153
consumers
   privacy, 185
   databases, 42
   First Virtual, 206
   marketing, 40
   mass marketing, 37
   complaints, 41
Copyright Clearance Center (Open Market), 154
CPM (cost per thousand), 36
cracking encryption, 59
credit (Netscape), 74
credit cards
   CheckFree, 163
   CyberCash, 216
   First Virtual, 136
   MasterCard, 159-160
   WWW site, 241
   online services, 65
   smartcards, 77
   Visa, 158
crime, 181
Crokett, Dr. Stephen (CyberCash), 116
cryptanalysis, 46

cryptography, 46
   Clipper Chip, 51
   private, 46
   public, 46
      nonrepudiation, 49
   RSA's FAQ, 58
   WWW site, 61
cryptology, 46
   DES (Digital Encryption Standard), 50
   digital signatures, 49
   Quadralay's Cryptography site, 48
CUCInternational (NetMarket), 173
CyberCash, 216
   address, 240
   banks, 114
   CheckFree, 123
   consumer accounts, 118
   credit cards, 216
   Crockett, Dr. Stephen, 116
   DES (Digital Encryption Standard), 118
   ecash, 121
   First National Bank of Omaha, 124
   Lynch, Daniel, 115
   Melton, William, 114
   partnerships, 122
   Secure Internet Payment Service, 216
   TNS (Transaction Network Services), 115
   vendors, 123
   Web Threads site, 121
   Webthreads site, 120
   Wells Fargo & Company, 124
   Wilson, Bruce, 116
   WWW site, 119, 240

D

damages (Mondex cards), 92
databases
   consumers, 42
   marketing, 44
Davidow, William, 19
Death of Money (book), 17
deciphering DES (Digital Encryption Standard), 50
demographics, 36
DES (Digital Encryption Standard), 50
   CyberCash, 118
   deciphering, 50

Index

DES source code, 51
DigiCash, 99, 103
  address, 240
  banks, 112
  Chaum, David (interview), 226
  Ecash, 104-106, 217
  electronic wallets, 107
  facility cards, 110
  smart cards, 108-110
  storefronts, 218
  WWW site, 240
DigiCash's Ecash home page, 107
digital cash
  banks, 178, 180
  CyberCash, 121
  DigiCash, 104
  interest, 179
  mutual funds, 188
  NetCash, 168
  private systems, 189
digital signatures, 49
  blind signatures, 101-102
directories (Yahoo), 31

E

E-cash, 79-80
  Matonis, Jon, 81
  Mondex, 90
  Proton Scheme, 93
  requirements, 82
e-mail, 27
  digital signatures, 49
  First Virtual (transactions), 203-204
  junk mail, 194
  marketing, 40
  message digests, 49
ECash, 106
  software, 217
ecash, *see* digital cash
EDI (Electronic Data Interchange), 98
EDS (First Virtual), 132
Electronic Payment Services
    (SmartCash), 241
Electronic wallets
  CheckFree, 123, 163
  DigiCash, 107
  see also wallets

encryption
  Clipper Chip, 51
  cracking, 59
  DES (Digital Encryption Standard), 50
  key escrow, 52, 55
  Lucifer, 50
  NetMarket, 173
  Netscape, 57
  PGP(Pretty Good Privacy), 54
  RC4 128-bit, 59
  RC4 40-bit, 58
  Telequip, 172
EPS (Smart Cash), 172
escrow key system, *see* key escrow
EShop (virtual mall), 70
Europay (iKP), 77

F

Facility cards (DigiCash), 110
FAQs (First Virtual), 133
Federal Express, 65
Federal Reserve, 176
  checks, 178
fees
  CommerceNet, 214
  InfoHaus, 201
fields (First Virtual), 205
Financial Services Technology
    Consortium, 162
Finger (First Virtual), 203
First Virtual Map site, 128
First National Bank of Omaha
    (CyberCash), 124
First Union, 156
  Open Market, 146
First Virtual, 125, 197
  address, 241
  advance payment, 202
  billing and collections, 131
  Borenstein, Nathaniel, 135
  buyers, 206
  clients, 136
  credit cards, 136
  e-mail, 203
  EDS (Electronic Data Systems), 132
  FAQs, 133
  fields, 205

Finger, 203
FV-API, 203
InfoHaus, 133, 199
    fees, 201
    WWW site, 200
InfoMerchants, 133
Internet Payment System, 200
NetIt, 137
payment mechanisms, 130
Rose, Marshall, 135
Rubin, Frank, 128
security, 128
selling, 197
setting up accounts, 197
SGCP(Simple Green Commerce Protocol), 204
Stefferrud, Einar, 134
Stein, Lee, 134
storefronts, 198
Telnet, 203
transactions, 128, 199-200
    automating, 204
    e-mail, 204
    Telnet, 205
virtual corporation, 133
WebIt, 137
WWW site, 241
Ford, Henry, 35
forgeries (checks), 178
freelancing, 20
Frontier Technologies (CyberCash), 122
FTP (File Transfer Protocol), 28
    CyberCash, 122
FV-APE (First Virtual), 203

G

gateways, 26
Ghosh, Shikhar (Open Market), 139
    interview, 237
Gifford, David (Open Market), 139
Gilder, George, 25
Gleason, Donald (Smart Card Enterprise), 234
Gopher, 29
Gore, Al (Clipper Chip), 57
government, 175
Green Card Lawyers, 26

H

Higgns, Graham, 84
Home Page for PC Flowers and Gifts, 68
host access, 26
hypertext, 30
    S-HTTP (Secure Hypertext Transfer Protocol), 76

I

IBM
    iKP, 76
    S-HTTP (Secure Hypertext Transfer Protocol), 77
ICS (Integerated Commerce Service), 144
IKP(IBM), 76
Industry (Open Market), 220
InfoHaus (First Virtual), 133, 199
    fees, 201
    WWW site, 200
InfoMerchants (First Virtual), 133
InterCon Systems (CyberCash), 122
Interest (ecash), 179
Internet, 21-22
    access, 26
    ARPANET, 22
    e-mail, 27
    FTP (File Transfer Protocol), 28
    gateways, 26
    Gopher, 29
    IP(Internet Protocol), 22
    ISDN lines, 27
    LANs (Local Area Networks), 23
    mainframes, 22
    marketing, 40
    modems, 27
    netiquette, 193
    NFSNET, 23
    NFSTNET, 25
    packet switching, 22
    PRNET, 25
    SATNET, 25
    self policing, 26
    TCP/IP, 23
    Telnet, 28
    Usenet, 29

Index

WAIS (Wide Area Information Server), 30
WWW, 30
Internet Society (ISOC), 26
InternetMCINavigator, 73
IP (Internet Protocol), 22
ISDNlines, 27
ISOC(Internet Society), 26

J-K

Jones, Tim (Mondex), 84
   interview, 221
junk mail, 194

kangaroo courts (Internet), 193
Kerberos software (NetCheque), 170
key escrow, 52
   encryption, 55
keys (Netscape), 60
Kurtzman, Joel, 17

L

labor (outsourcing), 20
LANs (Local Area Networks), 23
Lee, Roger (NetMarket), 174
Lexis-Nexis (Open Market), 150
Lucifer (encryption), 50
Lynch, Daniel (CyberCash), 115

M

mail order
   Federal Express, 65
   sales tax, 177
mainframes, 22
malls (virtual), 66
Malone, Michael, 19
marketing, 36
   consumers, 40
   CPM (cost per thousand), 36
   databases, 44
   demographics, 36
   e-mail, 40
   Internet, 40
   one-to-one relationships, 42
MarketplaceMCI, 71-73
   baseball, 71
   InternetMCINavigator, 73

mass marketing, 36
   choice, 37
mass production, 35
MasterCard, 159-160
   WWW site, 241
Matonis, Jon, 79
MCI
   address, 241
   WWW site, 241
Melton, William (CyberCash), 114
Membership (CommerceNet), 210, 213
Merchants (Open Market), 142
Message digests, 49
Metaverse, 1
Modems, 27
Mondex, 84
   address, 241
   banks, 97
   children, 89
   damaged cards, 92
   e-cash, 90
   EDI (Electronic Data Interchange), 98
   Jones, Tim, 221
   purchases, 91
   reverse engineering, 94
   security, 94, 96
   Swindon, 87
   transactions, 91
   unaccounted system, 93
   wallets, 89
   WWW site, 241
Mosaic (NetMarket), 173
Mutual funds, 188
National Science Foundation (NSF), 23
National Writers Union, 189
NetCash, 168
   address, 241
   NetCheque, 168
   WWW site, 241
NetCheque, 168
   Kerberos software, 170
   NETCOM(CyberCash), 122
   Netiquette, 193
NetIt (First Virtual), 137
NetMarket, 173
   address, 242
   CUC International, 173

InfoHaus (First Virtual), 133, 199
Lee, Roger, 174
Mosaic, 173
PGP(Pretty Good Privacy), 173
WWW site, 242
Netscape, 171
   address, 242
   Clark, Dr. James H., 171
   credit, 74
   encryption, 57
   keys, 60
   protocols, 171
   RC4 40-bit encryption, 59
   S-HTTP (Secure Hypertext Transfer Protocol), 76
   Secure Courier, 171
   security breach, 60
   Terisa Systems, 76
   WWW site, 242
network access, 26
newsgroups, 29
   charters, 193
   netiquette, 193
   threads, 29
newsreaders, 29
NFSNET, 23
NFSTNET, 25
nonrepudiation (public cryptography), 49
NSA (National Security Agency), 51
   Clipper Chip, 56

O

One-to-one relationships (marketing), 42
Online services
   credit cards, 65
   Prodigy (virtual mall), 66
Open Market, 138, 218
   address, 242
   Advance Publications, 151
   Bank One, 148, 150
   Boston Camera View, 138
   Community-Commerce, 147
   Condé Nast, 153
   Copyright Clearance Center, 154
   CyberCash, 122
   First Union, 146
   Ghosh, Shikhar, 139
      interview, 237
   Gifford, David, 139
   ICS(Integrated Commerce Service), 144
   industry, 220
   Lexis-Nexis, 150
   merchants, 142
   publishing, 218
   remote retailers, 219
   S-HTTP, 143
   servers, 143
   Small Business Advisor, 151
   Stewart, Lawrence, 139
   Time, Inc., 151
   transactions, 141
   Tribune Company, 151
   WWW site, 242
outsourcing labor, 20

P

packet switching (Internet), 22
payment schemes site (WWW), 64
payments (First Virtual), 130, 200-202
PC Flowers, 67
PDAs (Personal Data Assistants), 107
Personal assistants (virtual malls), 69
PGP(Pretty Good Privacy), 54
   NetMarket, 173
Premium Advantage (virtual mall), 70
Privacy, 184, 187
   blind signatures, 100
   Clipper Chip, 53
   publishing, 190
private cryptography, 46
private ecash systems, 189
private sector (Clipper Chip), 55
PRNET, 25
Prodigy
   PC Flowers, 67
   virtual mall, 66
production speed, 18-19
protocols
   Netscape, 171
   SSL (Secure Sockets Layer), 57
   TCP/IP, 23
Proton Scheme (e-cash), 93

Index

public cryptography, 46
   nonrepudiation, 49
publishing
   Open Market, 218
   privacy, 190
purchases (Mondex), 91

Q

Quadralay's Cryptography site, 48
Quarterdeck (CyberCash), 122

R

RC4 128-bit encryption, 59
RC4 40-bit encryption, 58
   Netscape, 59
redlining (banks), 179
registration (CommerceNet), 213
remote retailers (Open Market), 219
reverse engineering (Mondex), 94
Rolling Stones web site, 32
Rose, Marshall (First Virtual), 135
RSA's FAQ about cryptography, 58
Rubin, Frank (First Virtual), 128

S

S-HTTP (Secure Hypertext Transfer
     Protocol), 76
   Open Market, 143
sales tax, 177
SATNET, 25
Schutzer, Dan (Citibank), 230
scripts (First Virtual transactions), 204
search engines (WWW), 31
secrecy (Clipper Chip), 56
Secure Courier (Netscape), 171
Secure Internet Payment Service
     (CyberCash), 216
security
   blind signatures, 100
   First Virtual, 128
   Kerberos software, 170
   Mondex, 94, 96
   NetCash, 168
   Netscape, 57
     breach, 60
Security First Network Bank, 242

self policing (Internet), 26
selling (First Virtual), 197
servers (Open Market), 143
SGCP (Simple Green Commerce
     Protocol), 204
signatures (digital), 49
Skipjack (Clipper Chip), 53
Small Business Advisor (Open
     Market), 151
Small Businness Evangelist Program
     (CommerceNet), 215
Smart Card Enterprise (Donald
     Gleason), 234
smart cards, 77
   DigiCash, 108-110
   Financial Services Technology
     Consortium, 162
   Mondex, 84
Smart Cash (EPS), 172
sociological impact (digital cash), 183
software
   Clickshare, 191-193
   ECash, 217
   First Virtual, automating
     transactions, 204
   newsreaders, 29
speed (production), 18-19
SSL (Secure Sockets Layer), 57
Starter kit (CommerceNet), 213
Stefferuc, Einar (First Virtual), 134
Stein, Lee (First Virtual), 134
Stewart, Lawrence (Open Market), 139
storefronts
   CommerceNet, 214
   DigiCash, 218
   First Virtual, 198
Swindon (Mondex), 87

T

taxes ()sales, 177
TCP/IP, 23
Telequip, 172
Telnet, 28
   First Virtual, 203
     transactions, 205
Terisa Systems, 76
terminal access, 26

threads (newsgroups), 29
Time, Inc. (Open Market), 151
TNS (Transaction Network Services), 115
Toyota, 19
transactions, 188
   blind signatures, 101
   First Virtual, 128, 199-200
      automating, 204
      e-mail, 204
      Telnet, 205
   Mondex cards, 91
   NetCash, 168
   Open Market, 141
Tribune Company (Open Market), 151

U-V

unaccounted systems (Mondex), 93
Usenet, 29
   see also newsgroups
vendors (CyberCash), 123
ViaCrypt, 242
virtual corporation (First Virtual), 133
Virtual Corporation, The (book), 19
virtual malls, 66
   AT&T, 69
   CompuServe, 69
   eShop, 70
   MarketplaceMCI, 71
   personal assistants, 69
   Premium Advantage, 70
Visa, 158
   Visa's Home Page, 158
   WWW site, 242

W-Z

WAIS (Wide Area Information
      Server), 30
wallets
   CheckFree, 123
   DigiCash, 107
   Mondex, 89
Web Threads site (CyberCash), 121
WebIt (First Virtual), 137
Webthreads site (CyberCash), 120
Wells Fargo & Company
      (CyberCash), 124
Wilson, Bruce (CyberCash), 116

wiretaps (Clipper Chip), 53
writers, 189
   Clickshare, 191
   privacy, 190
WWW (World Wide Web), 30
   CheckFree site, 240
   CitiCorp site, 240
   CommerceNet site, 240
   cryptography site, 61
   CyberCash site, 119, 240
   DES source code site, 51
   DigiCash site, 240
   DigiCash's Ecash home page, 107
   First Virtual, 241
   First Virtual Map site, 128
   First Virtual site, 125
   Home Page for PC Flowers and
      Gifts, 68
   hypertext, 30
   InfoHaus, 200
   MCI site, 241
   Mondex, 241
   Mosaic (NetMarket), 173
   NetCash site, 241
   NetMarket site, 242
   Netscape, 171
      security breach, 60
      SSL(Secure Sockets Layer), 57
      WWW site, 242
   Open Market site, 242
   payment schemes site, 64
   Quadralay's Cryptography site, 48
   Rolling Stones, 32
   RSA's FAQ about Cryptography, 58
   search engines, 31
   Security First Network Bank site, 242
   ViaCrypt site, 242
   Visa site, 242
   Visa's Home Page, 158
   Webthreads site (CyberCash), 120
   Yahoo, 31

Yahoo (WWW), 31